Anti-Microbial Resistance in Global Perspective

Louise Ackers · Gavin Ackers-Johnson ·
Joanne Welsh · Daniel Kibombo · Samuel Opio

Anti-Microbial Resistance in Global Perspective

palgrave macmillan

Louise Ackers
Global Social Justice
University of Salford
Salford, UK

Joanne Welsh
University of Salford
Salford, UK

Samuel Opio
Pharmaceutical Society of Uganda
Kampala, Uganda

Gavin Ackers-Johnson
University of Salford
Salford, UK

Daniel Kibombo
Infectious Disease Institute
Kampala, Uganda

ISBN 978-3-030-62661-7   ISBN 978-3-030-62662-4   (eBook)
https://doi.org/10.1007/978-3-030-62662-4

This Palgrave Macmillan imprint is published by the registered company Springer Nature Switzerland AG
The registered company address is: Gewerbestrasse 11, 6330 Cham, Switzerland

We dedicate this book to the Future Generations of public health users in Uganda and Low- and Middle-Income Countries facing the impacts of Anti-Microbial Resistance.

FOREWORD

As the UK Special Envoy on AMR, I welcome Professor Louise Ackers' book, *Anti-Microbial Resistance in Global Perspective*. The resistance of bacteria to antibiotics is a growing threat to modern medicine and our ability to treat infections. It is estimated that more than 30,000 women die each year as a result of severe infections when giving birth and the majority of these deaths occur in LMICs. Sepsis accounts for around 10–15% of maternal mortality and the vast majority of these deaths could be prevented through the appropriate use and stewardship of antibiotics. Anti-microbial resistance is an increasingly serious threat to the gains made in health and development and for the attainment of the Sustainable Development Goals (SDGs) in particular, SDG 3.1 which calls for a reduction in preventable maternal deaths. We know that the situation will only get worse if, and when, the antibiotics that treat these infections become less effective. We need to protect and preserve the antibiotics for as long as possible to save the lives of millions not only for now but for the future generations.

This book provides us with invaluable empirical evidence on how we can promote the rationale use of antibiotics in LMICs to reduce the risk of pregnant women dying from a fatal drug resistant infection. Through effective infection prevention control and antibiotic prescribing based on laboratory results, Professor Acker shows us how rates of infection can be drastically reduced and more effectively treated. Using laboratory results as a basis for interventions is so important and leads to sustainable change. This was all achieved through a health partnership between volunteers in NHS Trusts and health institutions in the UK and their counterparts in Uganda and demonstrates the power of peer to peer support, in the form of mentoring and other less traditional approaches on formal training interventions, for effective behaviour change. We thank our health workers, in particular those involved in this partnership, whose great working is driving amazing progress in Uganda and beyond. With small and low-cost stewardship interventions we can make a significant impact on maternal mortality, and ultimately reduce the impact of antimicrobial resistance.

Professor Louise Ackers' work was funded by the Fleming Fund, a £265 million UK aid programme, through the Commonwealth Partnership on Antimicrobial Stewardship programme. The Fleming Fund supports LMICs to generate, use and share antimicrobial resistance data so they can optimise the use of antibiotics and reduce drug resistance. It is committed to improving understanding of antimicrobial resistance by raising awareness of antimicrobial stewardship in LMICs. The stewardship of antimicrobials currently available to us is vital in order to preserve the effectiveness of drugs for as long as possible and hence is a key component of our response to AMR.

Professor Dame Sally Davies
UK Special Envoy on AMR
London, UK

ACKNOWLEDGMENTS

The Maternal Sepsis Intervention was very much a team activity. We would like to thank all the members of the Project Team for their commitment and professionalism including; James Ackers-Johnson (Knowledge for Change), Allan Ndawula (K4C), Ibrahim Mugerwa (AMR National Taskforce), Saarah Niazi-Ali (Antibiotic Specialist Pharmacist, Tameside General Hospital, Manchester), Clare Liptrott (Pharmacist, University of Salford) and Allan Muwazae (Senior Pharmacist, FPRRH). Amrit Atwal (Anthony Nolan Haematology Pharmacist at University Hospitals Birmingham NHS Foundation Trust) was awarded a Chief Pharmaceutical Officer's Global Health Fellowship to work with the team. Chloe James (Microbiologist, University of Salford) supervised Gavin Ackers-Johnson's doctoral research and provided training on the interpretation of laboratory results in Uganda. Simon Sseguya (Pharmacist, FPRRH) worked closely with the team in supplies management and as a clinical pharmacist and joined colleagues in the UK in February 2020 for 4 months on a Commonwealth Professional Fellowship.

The Project owes special gratitude to the K4C clinical team on the ground including Dorothy Gashuga (K4C Lead Midwife); Rachel Namiiro (K4C Midwife), Barbra Kamara (K4C IPC Specialist Midwife) and Joe Odur (K4C Pharmacist). Dorothy, Rachel, Barbra and Joe have worked tirelessly on the wards for the whole project exuding professionalism and collegiality. They have also played a key role in data collection.

The work was supported by the presence of Professional Volunteers: Elizabeth Pearson, Rob Harper, Caroline Hammond, Amy Cleese, Jan Dubbink, Jemma Berwick and Harriet Earp.

Maaike Seekles provided support with data collation and analysis.

We would like to thank the Directors (past and present) of FPRRH,[1] Dr. Florence Tugumisirize and Dr. Alex Adaku and the Hospital Administrator, Mr. Louis Muhindo, for their active support with all aspects of the project.

We extend our gratitude to the health worker team on the Postnatal and Gynaecology wards whose commitment and expertise have been pivotal to project success. Special thanks go to the ward In-Charge, Mrs. Gladys Acayo, Hannah Kemiyondo, Irene Lubambula, Bridget Kyomuhendo and all the nurses and midwives.

This research was conducted during activities funded by the Department of Health and Social Care using UK aid funding and is managed by the Fleming Fund. The views expressed in this publication are those of the author(s) and not necessarily those of the Department of Health and Social Care.

---

[1] Fort Portal Regional Referral Hospital is known locally as Buhinga Hospital.

CONTENTS

CONTRIBUTORS

Gavin Ackers-Johnson University of Salford, Salford, UK

Louise Ackers Global Social Justice, University of Salford, Salford, UK

Daniel Kibombo Infectious Disease Institute, Kampala, Uganda

Samuel Opio Pharmaceutical Society of Uganda, Kampala, Uganda

Joanne Welsh University of Salford, Salford, UK

Acronyms

AMR	Antimicrobial Resistance
AMS	Antimicrobial Stewardship
API	Analytical Profile Index
AR	Action Research
BAP	Blood Agar Plates
CAI	Community Acquired Infection
CDC	Centre for Disease Control
CME	Continuing Medical Education
CPA	Commonwealth Pharmacists Association
CwPAMS	Commonwealth Partnerships for Antimicrobial Stewardship
DfID	Department for International Development
DHSC	Department of Health and Social Care
EMHSLU	Essential Medicines and Health Supplies List for Uganda
FPRRH	Fort Portal Regional Referral Hospital
GAP	Global Action Plan
GPPS	Global Point Prevalence Survey
HCAI	Health Care Acquired Infection
HIC	High Income Country
ICAN	Infection Control Africa Network
IDI	Infectious Diseases Institute
IPC	Infection Prevention Control
K4C	Knowledge for Change NGO
KHP	Kabarole Health Partnership
LMIC	Low- and Middle-Income Country
MCA	MacConkey Agar
MPDSR	Maternal and Perinatal Death Surveillance and Response

MSA	Mannitol Salt Agar
MSI	Maternal Sepsis Intervention
MTC	Medicines Therapeutic Committee
NAMRSC	National Antimicrobial Resistance Sub-Committee
NAP	National Action Plan
NICU	Neo-Natal Intensive Care Unit
NMS	National Medical Stores
NVIVO	NVivo is a Qualitative Data Analysis Computer Software Package Produced by QSR International
ODA	Official Development Assistance
PBS	Phosphate Buffered Saline
PNG	Post-Natal and Gynae
PPP	Public Private Partnership
PSU	Pharmaceutical Society Uganda
PYR	Pyrrolidonyl Aminopeptidase Test
SDG	Sustainable Development Goal
SSI	Surgical Site Infection
THET	Tropical Health Education Trust
UNHCR	United Nations High Commissioner for Refugees
UTI	Urinary Tract Infection
VfM	Value for Money
WASH	Water, Sanitation and Hygiene
WHO	World Health Organisation

List of Figures

LIST OF TABLES

CHAPTER 1

Introduction to Antimicrobial Resistance and the Maternal Sepsis Intervention

Abstract This chapter describes the threat to global health and security caused by the growing resistance of infectious organisms to antibiotics or antimicrobial resistance (AMR). Growing global connectivity ensures that AMR is a threat to us all wherever we are and with specific impacts on Low- and Middle-Income Countries (LMICs). The chapter outlines international responses to AMR including the Global Action Plan and the impact this has had on one LMIC; Uganda. It then introduces a recent UK funding call focused on improving the management of antibiotics or 'Antimicrobial Stewardship'.

Keywords Antimicrobial resistance · Antimicrobial stewardship · Infection prevention control · Universal health coverage · International development · Global health

© The Author(s) 2020
L. Ackers et al., *Anti-Microbial Resistance in Global Perspective*,
https://doi.org/10.1007/978-3-030-62662-4_1

ANTIMICROBIAL RESISTANCE; A POST-ANTIBIOTIC APOCALYPSE?[1]

Infection of the human body by an external vector, often a microorganism, has been a major cause of human illness and death. The development of antibiotics and their therapeutic use since the 1940s has played a critical role in the prevention and treatment of infectious disease. So much so, that there is a growing call for a shift in emphasis in global health, towards investment in non-communicable diseases such as heart disease, cancer or diabetes. Evidence of growing resistance of infectious microorganisms to commonly used antibiotics has raised serious concerns about the ability to continue to control infection and manage communicable disease. Microorganisms, including bacteria, can constantly mutate and resist the presence of antibiotics. This mutation is a naturally occurring process, but it is accelerated by the inappropriate use of antibiotics. Some of the routinely used antibiotics now have limited if any impact on common infections. Put simply, they no longer work. This has led to the development of newer antibiotics which are often more expensive for users and health systems, but many of these are now also compromised. The use of antibiotics by humans is a key factor shaping resistance; however, there is growing recognition of the impacts that the use of antibiotics in animal health—and recently in agriculture—has on human health (and vice versa). This concern is captured by the concept of 'One Health', which takes a holistic view of disease and health care across both animal and human sectors.

The declining effectiveness of antibiotics in the treatment of infection not only affects communicable disease; it also impacts our ability to treat conditions such as cancer, as invasive treatment using chemotherapy, for example, generates low levels of immunity leaving patients vulnerable to infection.

The World Health Organization (WHO) defines antimicrobial resistance (or AMR) as the process of resistance that occurs when microorganisms (such as bacteria, fungi, viruses, and parasites) change when they are exposed to antimicrobial drugs (such as antibiotics, antifungals, antivirals, antimalarials, and anthelmintics). Microorganisms that develop antimicrobial resistance are sometimes referred to as 'superbugs'. As a result

---

[1] Professor Dame Sally Davies, Chief Medical Officer England (2017) https://www.antibioticresearch.org.uk/chief-medical-officer-calls-new-action-antibiotic-resistance/.

of growing resistance, the medicines become ineffective and infections persist in the body, increasing the risk of spread to others.[2]

A UK Review of antimicrobial resistance chaired by Jim O'Neill describes AMR as a 'Terrible Scenario' posing a genuine economic and security threat (2016: 1). With the potential to reverse decades of medical progress in the treatment of communicable disease, AMR is one of the biggest threats to global health, food security and development in the world today. O'Neill reported that currently there are 700,000 global deaths a year related to AMR which is estimated to reach as high as 10 million by 2050 (O'Neill 2016). Whilst the numbers quoted were met with some scepticism due to the patchiness of data available, they give major cause for concern. Whilst human health is a priority, AMR will also have an inevitable impact on the global economy. The same study estimated that between 2015 and 2050 AMR could cost globally US$100 trillion and a study published by the World Bank estimated a reduction in global GDP of between 1.1 and 3.8 percentage points between 2017 and 2050 due to AMR (World Bank 2017).

AMR is a complex phenomenon. Science communication has failed to increase public awareness of AMR, its causes, and the need for behaviour change. A popular misconception amongst those people who have a limited understanding is that it is the individual consuming the antibiotics that become immune to their effects, rather than the microorganisms. This can lead to a level of complacency that we have the power, through our own actions, to protect ourselves from AMR by reducing our own use of antibiotics and saving them for when we really need them. Whilst this can lead to under-consumption and put some individuals at risk, it also discourages a more global, collective, consciousness.

A report on antibiotic resistance by the Ugandan National Academy of Sciences (2015) refers immediately to the impact of 'increasing global connectivity' resulting not only in the movement of people, animals and trade, but also 'the rapid transport of infectious agents and their antibiotic resistant genes' (2015: iii). Antibiotic-resistant genes do not respect international borders and spill-over to compromise global health. O'Neill's report identified several 'market failures' affecting AMR. These included externality (or spill-over) effects where action or lack of action in one location impacts others. The authors argue that the rise of AMR is a global

---

[2] https://www.who.int/en/news-room/fact-sheets/detail/antimicrobial-resistance.

problem that cannot be addressed materially without a critical mass of countries coming together to implement consistent solutions;

> Drug resistant infections spread very quickly; a person carrying a drug-resistant bacterium can fly across the world in a matter of hours. Even if out of pure self-interest, it may make sense for high-income countries to support these efforts in lower income settings. (2016: 65)

Grounded in understandable self-interest, high-income countries are increasingly aware and concerned about growing evidence of the relationship between global health threats and political and economic instability (Saha and Alleyne 2018). The experience of the global COVID-19 pandemic has triggered an overwhelming awareness of global interconnectivity. Human behaviour and antibiotic consumption practices (amongst other things) in low- and middle-income countries (LMICs) immediately impact public health in high-income countries (HICs). The converse is of far greater concern as the over-consumption of antibiotics in more affluent countries by humans, animals and plants leads to a reliance upon newer, more expensive antibiotics beyond the reach of many people in LMICs, dependent on impoverished public services.

There is also a more normative concern, articulated by international organisations such as the United Nations and the World Health Organization, and underpinned by ethical commitments to concepts of regional equality and individual equity. The Sustainable Development Goals (SDGs)[3] were launched by the United Nations in 2016. SDG3 captures the commitment to ensuring healthy lives and well-being for all. Although SDG3 refers specifically to the need to combat epidemics and communicable diseases, there is no specific reference to the impact of AMR. Objective 3.8 enshrines a growing commitment, in global health, to 'universal health coverage, including financial risk protection, access to quality essential health-care services and access to safe, effective, quality and affordable essential medicines and vaccines for all'. AMR presents a major, if silent, threat to the achievement of universal health coverage in LMICs.

---

[3] https://www.un.org/sustainabledevelopment/sustainable-development-goals/.

Addressing Antimicrobial Resistance on a Global Platform

Alongside surveillance of resistance patterns, there is an emerging consensus that efforts to contain the threat of AMR should focus on four key objectives. The first is to address the inappropriate use of antibiotics in both human and animal populations, with high usage being the predominant factor in the development of AMR. Secondly, the effective treatment of liquid waste and sewage to reduce the volumes of antibiotics being dispersed into the environment. Thirdly, through utilising effective infection prevention and control (IPC) strategies in conjunction with water, sanitation and hygiene (WASH) programmes, the need for antibiotic usage can be reduced. Finally, it is essential that effective, affordable, and regulated antibiotics are available to those that require them to avoid the self-prescription of uncertified alternatives. These objectives are most effective in combination with new diagnostic and treatment technologies (Rochford et al. 2018).

In May 2015, the World Health Assembly endorsed a Global Action Plan (GAP) to tackle antimicrobial resistance, including antibiotic resistance, the most urgent drug resistance trend.[4] To achieve this goal, the GAP set out five Strategic Objectives:

1. Improve awareness and understanding of antimicrobial resistance.
2. Strengthen knowledge through surveillance and research.
3. Reduce the incidence of infection.
4. Optimise the use of antimicrobial agents.
5. Develop the economic case for sustainable investment that takes account of the needs of all countries, and increase investment in new medicines, diagnostic tools, vaccines, and other interventions.

The intention was that the GAP would be aligned to the national context of each member country that would then define a National Action Plan for implementation purposes.

---

[4] https://www.who.int/antimicrobial-resistance/global-action-plan/en/.

The Ugandan Antimicrobial Resistance National Action Plan

The Ugandan National Action Plan was published in December 2018. Echoing the Global Action Plan (GAP), it takes a 'whole-of-society engagement' approach, embracing the *One Health* Agenda. This ensures that AMR can be targeted at every level be it related to humans, animals or the environment. To help provide structure and enhance collaboration, a multidisciplinary National Antimicrobial Resistance Sub-Committee (NAMRSC) was formed which comprises of representatives from multiple industries including fisheries and agriculture, water and waste management, human health, wildlife, science and research and professional societies. Through the utilisation of such sub-committees, it is hoped that the NAP will have a national, regional, and local impact. Mirroring GAP, Uganda's National Action Plan specifies five key strategic objectives (GoU 2018):

1. Awareness
2. Infection Prevention and Control
3. Stewardship
4. Surveillance
5. Research and Innovation

The Commonwealth Partnerships for Antimicrobial Stewardship (CwPAMS) Project Call

In 2018, a consortium of UK organisations including the UK Department of Health and Social Care (DHSC), the Commonwealth Pharmacists Association (CPA) and the Tropical Health and Education Trust (THET) came together under the umbrella known as the Commonwealth Partnerships for Antimicrobial Stewardship (CwPAMS)[5] to launch a call for applications for funding. This scheme was part of the Fleming Fund, a wider programme managed by the DHSC that aims to help LMICs tackle antimicrobial resistance. Uganda was listed as one of four target countries along with Tanzania, Ghana and Zambia.

[5] https://www.thet.org/our-work/grants/cwpams/.

This book traces the intervention process from initial design of an application that would qualify for funding under the CwPAMS call and the subsequent intervention that took place. Significant attention is given (in Chapter 2) to project design processes in international development and the impact of funding systems on this. This is unusual in publications which often regard the funding mechanism as 'house-keeping' and simply acknowledge funders. We discuss it here as it provides a powerful illustration of the tensions that arise in international development between the increasingly prescriptive demands of funders, understandings of best practice in intervention research and the ethics of global partnerships. Chapter 2 outlines the objectives of the funders and the impact these had on intervention design and research/evaluation methods. It explains the reasons for focusing the work on maternal sepsis, one of the major causes of maternal mortality and antibiotic consumption in LMICs and describes the context within which the Maternal Sepsis Intervention took place; in a Regional Referral Hospital in Western Uganda.

In a further break with tradition, Chapter 3 presents the key outcomes of the intervention outlining major improvements in maternal mortality, length of patient stays, readmission rates and hospital expenditure. These outcomes are presented at the outset to facilitate a focus on change processes which we regard as the key findings that will inform a scalable intervention model. Chapter 4 outlines a key component of change. Ultimately, the optimal way of reducing unnecessary antibiotic use is to reduce the incidence of infection and, particularly the very high infection rates (or 'adverse events') associated directly with medical intervention and hospitalisation. The concept of infection prevention control (IPC) embraces a myriad of mundane, housekeeping practices routinely performed by grassroots cadres that are readily overlooked in highly medicalised or pharmaceuticalised agendas. Just as housework or care has been described by feminists as a 'Labour of Love' (Oakley 1974); constituted by numerous small tasks constantly un-done by those we serve, so too is basic IPC. It is the neglected foundation undertaken by underpaid and disempowered staff.

Chapter 4 addresses some of the issues most commonly associated with IPC; hand hygiene, waste disposal and infrastructure. It then addresses wound management as an infection control issue. The emergence of wound management as a central focus in the Maternal Sepsis Intervention proved pivotal in shaping the pathway to antimicrobial stewardship.

Chapter 5 moves on to discuss how wound management formed the catalyst for laboratory engagement and the taking of samples for testing. The laboratory results then created the 'objective' evidence-base that facilitated and nurtured midwifery empowerment, task-shifting and multidisciplinary team working. Chapter 6 addresses the role that pharmacy has and could increasingly play in antimicrobial stewardship in Ugandan public hospitals. The existence of laboratory results has triggered the emergence of clinical pharmacy roles with pharmacists playing an active role in multi-disciplinary teams. Chapter 7 traces the impact of this on prescribing behaviour (and individual patient benefits) and, more widely, on procurement planning and hospital policies. Whilst celebrating the progress made and viability of the model, Chapter 7 describes the structural impact that accesses to antibiotics and IPC supplies has on the realisation of optimal change. Chapter 8 reflects on the relationship between the knowledge mobilisation processes that have contributed to behaviour change at an individual and organisational level. It critiques the traditional emphasis in international development on one-off, formal, foreign-led 'training' episodes and contrasts these with the more fluid, bi-lateral, approach to learning through co-working and mentoring; approaches that sit uneasily with the accounting methods favoured by funding bodies. Chapter 9 summarises the overall impacts of the Maternal Sepsis Intervention reflecting on the processes of capturing, sustaining, and spreading best practice in antimicrobial stewardship.

References

Government of Uganda. (2018). *Antimicrobial Resistance National Action Plan 2018–2023.*

Oakley, A. (1974). *The sociology of housework.* Oxford: Blackwell.

O'Neill, J. (2016). *The Review on Antimicrobial Resistance. 2016. Tackling drug-resistant infections globally: Final report and recommendations.* The Review on Antimicrobial Resistance, Chaired by Jim O'Neill. Report commissioned by the UK Prime Minister.

Rochford, C., Sridhar, D., Woods, N., Saleh, Z., Hartenstein, L., Ahlawat, H., et al. (2018). Global governance of antimicrobial resistance. *Lancet, 391*(10134), 1976–1978.

Saha, A., & Alleyne, G. (2018). Recognising noncommunicable disease as a global health security threat. *Bulletin of the World Health Organisation, 96,* 792–793.

Uganda National Academy of Sciences. (2015). *Antimicrobial resistance in Uganda: Situation analysis and recommendations.* Centre for Disease Dynamics, Economics & Policy. ISBN: 978-9970-424-10-8.

World Bank. (2017). *Drug-resistant infections: A threat to our economic future.* Washington, DC: World Bank.

CHAPTER 2

Autonomy, Evidence and Methods in Global Health

Abstract This chapter discusses the growing impact that funding bodies have on the design, delivery and evaluation of global health interventions with specific emphasis on the UK's **Commonwealth Partnerships for Antimicrobial Stewardship** (CwPAMS) funding programme. It explains the reasons for focusing the antimicrobial resistance intervention on maternal sepsis and describes the context within which the Maternal Sepsis Intervention took place; in a Regional Referral Hospital in Western Uganda.

Keywords Antimicrobial resistance · Antimicrobial stewardship · Infection Prevention Control · International development · Global health · Maternal sepsis · Pharmaceuticalisation · Ethnography

This chapter addresses a key question: how do we create a high quality, Fit-For-Purpose, evidence base for global health interventions that optimally combines 'change' objectives with the generation of credible, scientific evidence?

Evidence in International Development, as in many policy domains, has played two rather different roles. On the one hand, it concerns the quest for an evidence base to guide policy. On the other, it represents

© The Author(s) 2020   11
L. Ackers et al., *Anti-Microbial Resistance in Global Perspective*,
https://doi.org/10.1007/978-3-030-62662-4_2

a response to an increasingly cynical political environment, in an age of austerity, that questions the efficacy of public expenditure on Aid; the 'giant cashpoint in the sky'.[1] Evaluation, as the generation of knowledge, then merges with financial and political accountability. Moyo captures this concern poignantly when she argues that Aid is malignant, and evaluation has contributed to a smoke screen:

> In nearly all cases, short-term aid evaluations give the erroneous impression of aid's success [...] The notion that aid can alleviate systemic poverty, and has done so, is a myth. (2010: xix)

Rajkotia (2018) makes a similar point referring to the enormous pressure on global health institutions and the foreign aid 'industry' to achieve targets. This he suggests leads to a tendency to 'embellish' reporting and, worse still, fabricate or overattribute achievements (p. 1). Storeng and Palmer (2019) detail how the pressure on donors to be seen to deliver on investments can contribute to serious challenges to researcher independence and even censorship of results. This, they argue, contributes to 'tick-box evaluations designed to please donors' (p. 185).

The emphasis on evaluation rather than research in funding calls is indicative of this merging of two rather uncomfortable bedfellows; financial accountability and knowledge. It lies behind the emergence of an entirely new cadre of impact assessment 'experts' and evaluators typically juggling identities as project managers/evaluators. The link with accountability mechanisms has, perhaps unintentionally, centre-staged a specific approach to evaluation that has been the subject of substantial critique for over 50 years in academic research. The term 'paradigm' is often used to describe the domination of a specific way of thinking that shapes attitudes and behaviour. And the paradigm we are referring to here has been known as 'positivism'.[2] Positivist methods—and the emphasis on measurable (quantitative) outcomes or 'ac/counting'—may meet the

---

[1] This was a comment made by UK Prime Minster Boris Johnson on 16 June 2020 to support the decision to merge the Department for International Development with the Foreign Office. https://www.bbc.co.uk/news/uk-politics-53062858.

[2] Put very simply positivism is an approach developed in the natural sciences which is based on the idea that facts exist (and can be mathematically verified) through scientific experimentation. It is also associated with deductive approaches and theory or hypothesis-testing. According to this approach, a researcher starts with a theory or hypothesis about a phenomenon or social problem and designs experiments to test that. The emphasis on

needs for financial accountability (or Value For Money: VfM). But is it the best approach to knowledge generation and transfer in health systems research?

Bridget Somekh traces the parallel development of increasingly radical social science theories and a policy context (the politics of sponsored research) that has, '*moved in the other direction and is ideologically framed now in more totalitarian assumptions of traditional research practices than was the case in the 1970s and 1980s*' (2006: 5). Somekh goes on to critique the positivist paradigm and, with specific reference to research on education systems in the UK, explains how this has led to an emphasis on technical solutions:

> ... we are locked in unrealistic assumptions of the application of natural science research methods to social situations; there is a belief in a process of incremental knowledge building to construct a technology of definite [educational] solutions for generalised application across contexts. (p. 5)

Harding makes a similar point in the context of international development critiquing not only the emphasis on training (a point we return to) but also the underlying methods and conclusions they infer:

> The transfer of Western scientific rationality and technical expertise from the West to "the rest" had always been the "motor" of modernisation theory and now drives development policy. However, many of the assumptions about women and poor people in the Global South - were false. (2015: 152)

The point about generalisability is important in the current context. One of the reasons for the emphasis on 'measurable outcomes' amongst international organisations, such as UKAid or the WHO lies in the perceived need to aggregate outcomes from diverse interventions to demonstrate cumulative impact (and benchmark change over time). How does an organisation such as the UK's Department for International Development (DfID), with a budget of £14 billion and under huge public scrutiny, capture impacts across a plethora of interventions spanning the scope of the Sustainable Development Goals (from health, gender equality,

---

'facts' (as opposed to knowledge) sees the data collector as necessarily external to this process—and as a potential pollutant.

economic growth etc.)? How do we compare the efficacy of a project on cervical cancer screening in Malawi with one on childhood disability in Sudan?

It would be hard to question the merits of this goal and understand the resort to metrics. But will those metrics become so generic that we focus on the measurable at the expense of the meaningful?

This is often the challenge facing not only end of the line project implementors and researchers but also intermediary funding organisations.[3]

Quick fix technical approaches based on highly unreliable and 'sanitised' secondary data are more likely to caricature reality than they are to capture genuine change and the costs associated with that. Harding, in a seminal feminist critique of research methods, refers to the use of evaluators in such situations as 'fast guns for hire' who can relay one version of events about the world 'ready-made for reporting' [but] without listening to women's accounts' (1991: 158). With reference to the claim to greater objectivity that underpins the positivist approach, Harding makes a very critical and relevant assertion:

> Paradoxically, the more "scientific" social research becomes, the less objective it becomes. (1991: 140)

The concept of objectivity is associated with notions of bias; it is based on the idea that there is a single truth that exists outside of any investigation, and the job of the researcher or evaluator is to avoid contamination of the data (facts). The concept of subjectivity, on the other hand, conveys the idea that a researcher is a person who interacts with the world and the people they are studying and their values and experiences inevitably shape the 'data' they generate. People have values that impact the way they see the world and influence 'truth' claims. Stephen Jay Gould points to the fundamental subjectivity of science:

> Science, since people must do it, is a socially embedded activity. It progresses by hunch, vision, and intuition [...] the most creative theories

---

[3] This aggregation process has many 'layers'. At program level, the CwPAMS can be viewed as an intermediary organisation reporting to higher level funders. In that respect, the first phase aggregation process seeks to align projects within a broadly similar framework (described below) albeit in very different contexts.

are often imaginative visions imposed upon facts; the source of imagination is also strongly cultural. (1981: 303)

These are complex philosophical concepts and we do not attempt to explore them in depth here; rather to illustrate their impact on the evidence base guiding international development in general and the conditions framing our antimicrobial resistance project in Uganda. In a field so explicitly and necessarily focused on change and with very powerful normative drivers (or value commitments) to universal health coverage, gender equality and empowerment—the approach to data and objectivity associated with positivism is simply untenable. These substantive value commitments that applicants for international development funding are required to (and should) align themselves to represent an immediate challenge to conventional ideas of objectivity and the 'd' words identified as synonyms of objectivity: disinterest; dispassion; detachment.

Harding argues that, '*It is a mistake to assume that research shaped by social values and interests invariably will be empirically unreliable. Maximal objectivity and a commitment to a more democratic organisation of the research process need not conflict... they can often enhance each other*' (2015: 151).

The concept of partnership working embraced in the Sustainable Development Goals and echoed in UKAid policies also presents serious challenges and lies in genuine and significant tension with the deductive principles underlying positivism. This is particularly problematic with short term, 'hit-the-ground-running' approaches to funding where applicants are required to present a highly specified 'theory of change'[4] at application stage. Even where scoping work[5] has been undertaken, outlining a theory of change at the start of a complex project on antimicrobial resistance is precisely the kind of deductive approach that fails to meet the principles of partnership outlined in SDG 17. The Health Partnership Scheme managed by the Tropical Health and Education Trust has developed principles of partnership that all applicants are required

---

[4] The 'theory of change' approach whilst echoing the language of academic research uses the concept of theory in a quite specific way (see Vogel 2012).

[5] Scoping visits are often of 1–2 weeks' duration and, at best, help to establish teams but not to understand context in any meaningful way.

to align themselves to.[6] These include 'listening to one another; cultivating trust; proactively adapting to change and aligning interventions to national planning' and suggest the need for a far more iterative, responsive and relational approach to both intervention and research than is possible within a positivist, theory-testing framework. The ideas of listening to each other resonates with Harding's description of research as an *'affirmation of ordinary life'* (1991: 158).

We would argue that the specific combination of change objectives with evaluation (research) predicated in most Official Development Assistance (ODA) work are best captured through the principles of Action Research. Action research is typically associated with more inductive approaches to theory generation; rather than starting with theories and trying to test them (somewhere else) 'in the field', action research is driven by contextual dynamics. Although researchers and actors will come to projects with prior knowledge (and in that sense cannot approach any social situation with an entirely blank slate) theory generation in projects will emerge informed by the context, ongoing review of other work and through active inter-personal relationships (partnership engagement). Somekh describes, '… *the ways in which social science researchers use action research methodology to overcome the limitations of traditional methodologies when researching changing situations. Action research combines research into substantive issues […] with research into the process of development in order to deepen understanding of the enablers of, and barriers to, change. It is a means whereby research can become systematic intervention, going beyond describing, analysing, and theorising social practices to working in partnership with participants to reconstruct and transform those practices. It promotes equality between researchers from outside the site of practice and practitioner-researchers from inside, working together with the aspiration to carry out research as professionals, with skilful and reflexive methods and ethical sensitivity'* (2006: 1).

Perhaps, with the exception of doctoral research, most research is conducted in partnership with funding bodies who have their own objectives. This inevitably shapes our ability to pursue the 'sociological imagination' (Wright-Mills 1959) and achieve optimal reflexivity. Harding acknowledges the need for compromise and recognition that other values will enter the negotiation process with funding bodies and experts:

[6] https://www.thet.org/principles-of-partnership/.

Scientists must balance their interests with those of their funders and sponsors even if one thinks they don't do so as vigorously as the anti-authoritarian and citizens science movements have been demanding. Scientists also negotiate routinely with experts of other fields with whom their research requires collaboration [...] negotiating such relations is what social life is about including the social life of science. (2015: 167)

Whilst we accept the need for balance, the result of this blurring of evidence with accountability has been for increasingly prescriptive funding calls to reduce project autonomy. This limits the scope for critical reflexivity and meaningful contextualisation. There is also a serious risk that high-level project management, by increasingly 'engaged' funding bodies, becomes a strait jacket with outcome measures framing interventions rather than permitting iterative, intelligent, approaches to influence outcomes. In effect, the tail may be increasingly wagging the dog to the detriment of knowledge and social change.

The specific nature of the funding stream as described above reflects the values of the parties involved, their approach to international engagement in general and AMR in particular. These have shaped the way the Maternal Sepsis Intervention has evolved often through 'creative tension'. We hope that by being explicit about this we can contribute to the co-production of more effective approaches to improve the evidence-base in global health. Achieving optimal objectivity in social research demands humility, honesty and the exercise of caution when making evidence claims. This requires openness about the objectives of funders, the normative underpinnings of these and the impacts these have on project design, implementation and outcomes.

The following section outlines the Call for Funding and the objectives and approaches applicants were invited to align themselves in order to position themselves for funding. The detail is presented here as an illustration of the level of complexity and prescription that shapes many, if not most, international development programs.

The Commonwealth Partnerships for Antimicrobial Stewardship (CwPAMS) Programme

The Commonwealth Partnerships for Antimicrobial Stewardship (CwPAMS) is a partnership involving three very different organisations: The UK Department of Health and Social Care, the Commonwealth Pharmacists Association and the Tropical Health and Education Trust. CwPAMS was awarded £1.3 million as part of a, 'wider commitment by the UK Government to spend up to £265 million of UK aid to support LMICs to enhance their surveillance of AMR by 2021'.[7] The Department of Health and Social Care manages this ambitious UKAid programme through the Fleming Fund. Although the Fleming Fund takes an holistic approach, there is a strong underlying emphasis on capturing 'surveillance' data to show patterns of resistance to antimicrobials (or which antibiotics are no longer effective in fighting infections). Despite the very serious and immediate threat that AMR poses to global health, there is still very little understanding of international patterns of resistance especially in LMICs where the capacity to generate and use surveillance data is particularly weak and uneven. Ultimately capturing resistance patterns, globally, represents the best approach to assessing the phenomenon of AMR; its responsiveness to environments (such as COVID-19) and interventions designed to contain it. This underlying emphasis on surveillance of resistance is expressed quite succinctly:

> The aim of the Fleming Fund is to get data relevant to AMR in the hands of decision makers. We want to support countries generating the data they need to inform policies and practices which will optimise the use of antimicrobial medicines.[8]

The Fleming Fund programme is made up of Country and Regional Grants and a Fellowships programme (administered by the Management Agent Mott MacDonald) and a variety of Global Projects managed directly by DHSC of which CwPAMs is one. The parent Fleming Fund includes extensive independent evaluation with an emphasis on:

---

[7] https://www.thet.org/our-work/grants/cwpams/.

[8] https://www.flemingfund.org/about-us/.

- how much the quantity and/or quality of data on AMR at country level has increased, and to what extent the Fleming Fund has contributed to this increase
- to what extent the Fleming Fund's investments have been aligned with other relevant investments at country level
- how sustainable the country level data quantity and/or quality increase is likely to be
- whether improved AMR data has influenced (a) changes in national policies/regulations, and/or (b) changes in practice and attitudes in each country
- how much the quality of data shared and reported internationally has improved, and whether the Fleming Fund has contributed to this
- whether the Fleming Fund's investments at country level offer value for money

The Tropical Health and Education Trust (THET) is a UK-registered charity focused on health system strengthening in LMICs through, 'training and educating health workers in Africa and Asia, working in partnership with organisations and volunteers from across the UK'.[9] THET was appointed as operational partner for the CwPAMS programme and allocated the funds through its established and prestigious 'Health Partnership' model. Twelve new and established Health Partnerships across four African countries (Ghana, Uganda, Tanzania, and Zambia) shared £600,000 of the £1.3 million in direct project funding.[10] Through, 'regular short-term visits' the partnerships were designed to, 'leverage the expertise of UK health institutions and technical experts to strengthen the capacity of the national health workforce and institutions to address **predefined** AMR challenges'.

The Commonwealth Pharmacists Association (CPA), on the other hand, is focused on Commonwealth countries and on improving the quality of pharmacy practice.[11] The CPA have acted as a technical partner

---

[9] https://www.thet.org/about-us/what-we-do/.

[10] The CwPAMS programme is subject to an additional evaluation managed by THET. In terms of VfM measures, the outcomes associated with the 12 projects sharing the £600,000 would be expected to demonstrate VfM for the full £1.3 million placing quite significant pressure on individual projects.

[11] https://commonwealthpharmacy.org/.

to THET on CwPAMS. Their role includes carrying out a scoping analysis, developing or providing assistance in the development of AMS resources, supporting grant holders with bespoke advice in AMS, and developing and analysing programme-level AMS reporting tools and data.

The predefined challenges referred to above were aligned to three of the Fleming Fund Objectives, with a specific focus on antimicrobial stewardship and the *use* of antimicrobials.[12]

Fleming Fund Priorities Identified by CwPAMS

- Support the development of National Action Plans for AMR
- **Developing and supporting the implementation of protocols and guidance for AMR surveillance and antimicrobial use.**
- Building laboratory capacity for diagnosis
- Collecting drug resistance data
- Enabling the sharing of drug resistance data locally, regionally, and internationally
- **Collating and analysing data on the scale and use of antimicrobial medicines**
- **Advocating for the application of data to promote the rational use of antimicrobials**
- Shaping a sustainable system for AMR surveillance and data sharing
- Supporting fellowships to provide strong national leadership in addressing AMR

The emphasis is on quite programmatic features with a strong assumption that LMICs need support in developing protocols and implementing these. Secondly, an emphasis on data and specifically antimicrobial consumption data. Finally, a strong normative assumption that this data will form the basis of effective advocacy for more rational use of antimicrobials. The Guidance to potential applicants added further complexity identifying three 'Themes'; with a *requirement* for each applicant to address Themes 1 and 2 (but not 3):

---

[12] Antimicrobials is a wider term referring to medicines that are active against microbes in general—including fungal and viral infections. The project has focused on bacterial infections and the use of antibiotics to treat those.

CwPAMS 'Themes'

1. Antimicrobial stewardship, including surveillance – requirement!
2. Antimicrobial pharmacy expertise and capacity – requirement!
3. Infection Prevention Control

Theme 1 reinforces the emphasis on stewardship recognising the relationships between stewardship (use of antimicrobials) and surveillance (of resistance patterns). Theme 2 adds a specific twist on this immediately identifying the role of pharmacy as the key discipline in antimicrobial use. The Guidelines then specified several outcomes:

CwPAMS Outcome Specifications

- Institutions and workforce demonstrate improved knowledge and practice related to AMS prescribing practice and IPC
- Evidence of effective AMR interventions, with standardised tools and guidance being used by local institutions and/or national partners
- NHS staff demonstrate improved leadership skills and a better understanding of the global context of AMR in their work

It is interesting to note that, although IPC was not a required 'theme', outcomes in relation to IPC were listed.[13] Outcome 1 anticipates both a knowledge premium (the ability to evidence **knowledge acquisition** amongst LMIC health workers) and improved practice (**knowledge application** or utilisation) with a very specific emphasis on '*prescribing* practice'. Echoing the Fleming Fund Objectives, Outcome 2 specifies a focus on the development of **standardised tools.** The emphasis on 'standardised' tools stands in some tension with the commitment to ensure contextual compatibility which lies at the heart of effective implementation. Whilst alignment with national or even international standards and explicit recognition of where such alignment is not possible is important, a 'one-size fits all' approach to something as complex as AMR is unlikely to be successful.

---

[13] CwPAMS hoped to embrace IPC through the programme as a whole but not necessarily in every project.

Outcome 3 draws on the Department of Health and Social Care's commitment to building the **expertise of the UK health workforce** (through volunteering). This introduces a welcome bilateral component to mutual learning but infers a difference in the type of knowledge to be acquired by the UK nationals and LMIC health workers. Whilst the knowledge premium for LMIC staff was anticipated in prescribing practice and IPC; the emphasis in UK volunteer learning was on leadership skills and an understanding of AMR in global context.

Later in the same guidance document Project Outcomes are re-stated but this time with a new emphasis on *microbiology data* which was not highlighted in the Fleming Fund Objectives. And IPC is very much present in this 'expectation'.

OUTCOMES EXPECTED FROM PROJECTS IN THE CwPAMS CALL[14]

What Outcomes Are Expected from You Through CwPAMS?

Partnerships should strengthen workforce in:

- Antimicrobial prescribing practice
- Use of microbiology data to inform decision making
- Infection Prevention Control
- Antimicrobial stewardship including surveillance of antimicrobial use

A further set of 'Scoping Requirements' placed a strong emphasis on 'antimicrobial consumption and behavioural drivers of inappropriate use'. This emphasis on antibiotic consumption and the behaviour of individual prescribers could be interpreted as falling within what Denyer Willis and Chandler (2018) characterise as a 'pharmaceuticalisation' model centred on pharmaceutical distribution and individual behaviour change. There is a strong emphasis in the 'Scoping Requirements' on the role of clinical pharmacy:

---

[14] These outputs were again expected across the programme and not necessarily within each project.

Project Activities should, '*Build on initiatives in the 4 countries for knowledge transfer and bidirectional learning to develop AMS as part of the **clinical pharmacy** role.*'

Importantly, the final 'Scoping Requirement' specifies a particular leadership model:

> Multidisciplinary team **led/co-led by pharmacists** that model best practice of multi-disciplinary working, especially nurses, pharmacists and doctors working equally.

Behaviour change was also high on the CwPAMS agenda with an emphasis on individual behaviour change as conceptualised in the Com-B Framework approach (Michie et al. 2011). And behaviour change scientists were enabled to join some of the funded projects to support implementation of this approach (Fig. 2.1).[15]

The approach should consider....
What are the behavioural barriers and drivers?

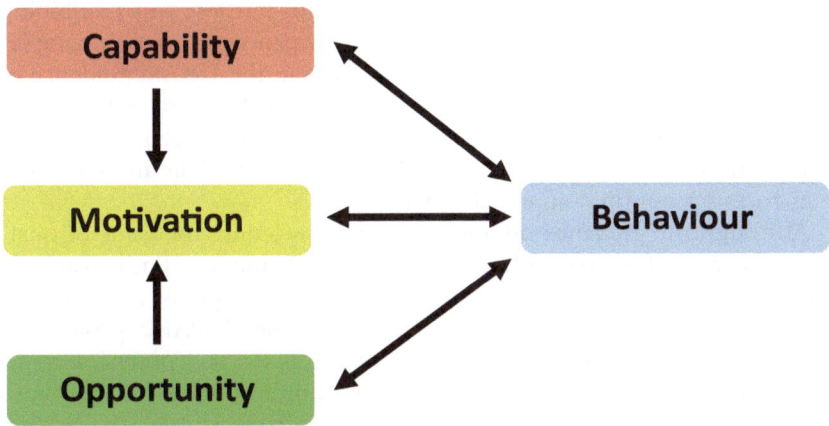

Fig. 2.1 Behaviour change theory in the CwPAMS programme (*Source* Adapted from Commonwealth Partnerships for Antimicrobial Stewardship Call for Applications Webinar)

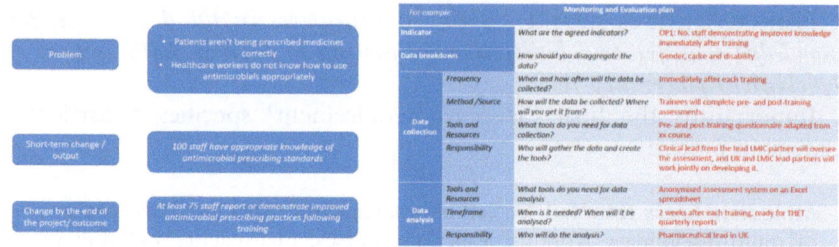

Fig. 2.2 Guidance on assessment of training (*Source* Commonwealth Partnerships for Antimicrobial Stewardship Call for Applications Webinar)

The Guidance also presented applicants with possible models of intervention and associated evaluation mechanisms with a powerful emphasis on 'measurable' outcomes. Figure 2.2 illustrate the anticipated/preferred approach and the emphasis on pre and post 'training' assessments.

It is important to note that the diagram presented above was for 'guidance' purposes only. However, successful applicants are required to complete a 'logframe' reporting template. UKAid has produced guidance on log frames and describes them as a multi-purpose tool combining 'regular project monitoring with annual review processes, project completion reports and evaluation'.[16] The description in the UKAid guidance of the 'results chain' as a 'logical (linear)' tool and subsequent reference to 'objective' measurement; the necessity of baseline data, attribution, and verification all point in the direction of what Denyer Willis and Chandler refer to as a 'counting' approach (2019: 2).

Although the guidance on log frames acknowledges the value of qualitative data it does so within an implicitly deductive, linear-planning perspective. The logic of the logframe methodology is expressed in Table 2.1 designed to capture the outcomes of the CwPAMS project.

Securing funding is a highly competitive process and potential applicants would ignore the guidance offered by funding bodies at their peril (Storeng and Palmer 2019). It is within this context that the project team designed their application for funding. To put the ambitious goals

---

[15] It is important to note that the COM-B Framework is only one approach to behaviour change theory (Ackers and Ackers-Johnson 2016).

[16] https://www.ukaiddirect.org/wp-content/uploads/2016/04/UKAD-Guidance-Log frames.pdf.

Table 2.1   Excerpt from CwPAMS logframe

Output 1	Output indicator 1.1
LMIC healthcare workforce strengthened in areas of AMS and antimicrobial prescribing practice	No. of LMIC healthcare staff trained in AMS, antimicrobial prescribing practise and consumption surveillance (based on WHO competency framework)
	Output indicator 1.2
	No. of LMIC healthcare staff trained and tested demonstrating improved knowledge after training
	Output indicator 1.3
	No. and % of LMIC healthcare staff able to demonstrate how to practise their new knowledge

Source Commonwealth Partnerships for Antimicrobial Stewardship Call for Applications Webinar

in context, the scheme launched on October 31st, 2018 with a submission deadline of January 4th, 2019. Grants were due to commence in February 2019. In practice, funding[17] was allocated in April 2019 with completion due on April 30th, 2020.

Whilst the value positions of funders will steer project design, the project team may be heavily influenced by recent research findings or alignment with their own partnership objectives and expertise. Research and interventions are often cumulative building on pre-existing work and relationships. Indeed, continuity is listed as one of THET's principles of partnership. The Kabarole Health Partnership (KHP), by way of example, had developed a strong area of expertise and associated relationships in maternal health and were acutely aware of the mortality associated with maternal sepsis. This introduces yet more complexity and 'steer' into the project planning process.

---

[17] 12 projects were awarded between £30,000 and £75,000 each; the MSI received £60,000.

WHY MATERNAL SEPSIS?

A recent review of research on antibiotic stewardship (Cox et al. 2017) found limited evidence of effective and feasible stewardship interventions in (LMICs) and, where examples of effective interventions were identified, emphasized the essential need for contextualization. From a hospital management and health worker perspective, outcomes focused on stewardship and antibiotic consumption do not immediately align with urgent and tangible service priorities. A key priority for Fort Portal Regional Referral Hospital[18] in 2018 was to reduce maternal mortality. As a Health Partnership, we were acutely aware of this priority and need.

Data from the Ministry of Health's most recent analysis of maternal and perinatal deaths (MOH 2019) indicate that, in the Financial Year 2018/2019, there were 1,180,321 deliveries in health facilities and 1083 maternal deaths. The majority of these deaths were reported from General (472) and Regional Referral Hospitals (334) and, of these, Fort Portal Regional Referral Hospital reported the second highest maternal mortality rate (Table 2.2).

According to this report, obstetric haemorrhage remains the leading cause of maternal deaths in Uganda accounting for 46% of all maternal deaths reported, followed by Infections/Anaemia/HIV & other conditions not related to Pregnancy (13%) and hypertensive disorders (11%). The 2019 MPDSR Report found that:

> Institutional maternal mortality ratios are highest at the regional and national referral hospitals (RRHs) (382/100,000 deliveries). This could be as a result of late and critical referrals from lower facilities, over-stretched resources (human, financial, equipment), inadequate essential supplies like blood and lifesaving commodities and the delays to access services at the referral sites. The institutions also received high numbers of patients in critical (near death) conditions. (2019: 24)

Ngonzi et al.'s study (2016) in Mbarara Hospital, Uganda reports puerperal sepsis[19] as the most frequent cause of maternal mortality responsible for 30.9% deaths as compared to obstetric haemorrhage (at 21.6%). Most cases of sepsis following childbirth can be characterised as Surgical Site

---

[18] FPRRH is often referred to locally as 'Buhinga Hospital'.

[19] Relating to the period up to 6 weeks after childbirth.

Table 2.2 Notification, reporting and reviews of maternal deaths at regional and national referral hospitals

	# Deliveries	# Maternal Deaths (HMIS 105)	IMMR/ 100,000 Deliveries.	# MDs Notified (WEP)	% MD.s Notified	# MD Reviews (Submitted at MoH)	% MD. Reviews
Arua RRH	7,625	15	196.7	11	73.3%	0	0.0%
Fort Portal RRH	7,600	48	631.6	27	56.3%	18	37.5%
Gulu RRH	3,257	4	122.8	3	75.0%	4	100.0%
Hoima RRH	8,213	65	791.4	54	83.1%	39	60.0%
Jinja RRH	6,535	11	168.3	12	109.1%	12	109.1%
Kabale RRH	3,613	4	110.7	2	50.0%	0	0.0%
Lira RRH	4,875	16	328.2	16	100.0%	6	37.5%
Masaka RRH	9,832	36	366.2	8	22.2%	16	44.4%
Mbale RRH	7,773	36	463.1	61	169.4%	11	30.6%
Mbarara RRH	9,189	42	457.1	26	61.9%	23	54.8%
Moroto RRH	808	2	247.5	-	0.0%	0	0.0%
Mubende RRH	5,328	29	544.3	14	48.3%	13	44.8%
Naguru Hosp-China Uganda Friendship	9,183	11	119.8	11	100.0%	7	63.6%
Soroti RRH	3,515	15	426.7	13	86.7%	15	100.0%
Mulago - Kawempe National RH	24,115	34	141.0	0	0%	6	17.6%
Grand Total	87,346	368	330.2	258	70%	164	46.2%

Source The National Annual Maternal and Perinatal Death Surveillance and Response (MPDSR) Report FY 2018/2019, Ministry of Health Uganda (September 2019)

Infections arising as a direct result of medical intervention (caesarean-section). A project focus on stewardship in relation to maternal sepsis had numerous attractions:

1. Sepsis is a major cause of maternal mortality
2. Sepsis is highly preventable through improved Infection Prevention-Control
3. Sepsis is associated with very high antibiotic consumption

The proposal is built on pre-existing work by the Kabarole Health Partnership (KHP). KHP engages a range of stakeholders in Uganda and the UK including Knowledge For Change (K4C), an NGO registered in the UK and Uganda; the Universities of Salford and Mountains of the Moon; Kabarole Health District and Fort Portal Regional Referral

Hospital (FPRRH). K4C functions in an operational role implementing projects on the ground and has a strong local presence in facilities. K4C's continuous presence on the ground and growing recognition of our principles of partnership and co-presence has led to strong relationships. In order to extend existing relationships to embrace pharmacy and ensure active engagement with the Ugandan National Action Planning process, the partnership expanded to include joint leadership with the Secretary General of the Pharmaceutical Society of Uganda (PSU) and the NAP Team. We also brought in pharmacy expertise from the University of Salford, linking directly to the non-medical prescribing programme[20] and the lead AMR pharmacist in Tameside and Glossop Hospital Trust.

Based on prior experience, the team proposed a 'Complex Intervention,' whole-systems, approach that built on pre-existing knowledge and explicitly allowed for flexibility in response to local contextual dynamics. McCormack describes the dual focus of action research (AR), combining the quest for new knowledge with the goal of achieving social change, as a reason for considering AR in the implementation of complex interventions (2015: 300). He further argues that much of the complexity in complex interventions arises from the context within which any evidence is to be implemented and, citing Bates, suggests that, 'nothing exists, and therefore can be understood, in isolation from its context' (2014: 3). The following section outlines the context within which the MSI developed.

STUDY CONTEXT: THE POST-NATAL AND GYNAECOLOGY WARDS AT FPRRH

The Kabarole Health Partnership (KHP) had an active presence on the maternity wards at Fort Portal Regional Referral Hospital. There are two postnatal wards in FPRRH. One is designated for women who have had vaginal births, the other for those who have had caesarean sections. Our focus was on the women on the caesarean section post-natal ward. If these women recovering from c-sections become unwell, they are transferred to the adjoining gynaecology ward or High Dependency Unit. Suspected

---

[20] Non-medical prescribing is possibly one of the most inappropriately named task-shifting programs. In practice, non-medical prescribing means prescribing by non-medical cadres and is actively supported in the UK through an accredited training programme; https://www.rcn.org.uk/get-help/rcn-advice/non-medical-prescribers. No such programmes exist in Uganda.

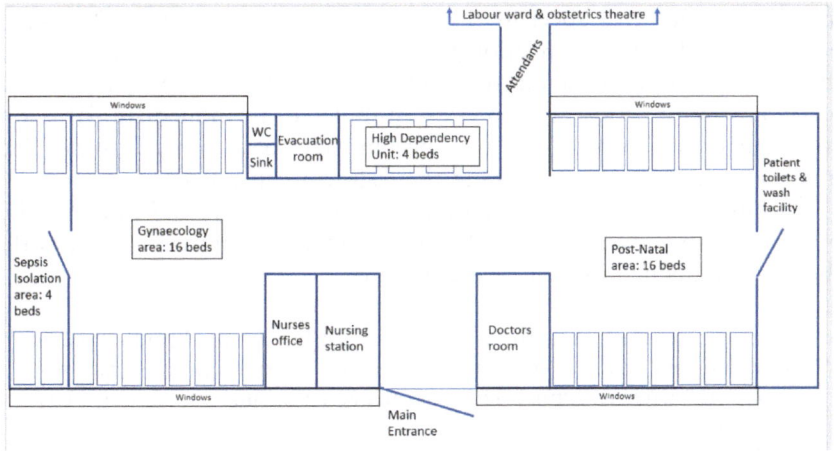

Fig. 2.3 Site plan of the post-natal and gynaecology wards at FPRRH

sepsis cases are managed in an isolation area at the far end of the ward as indicated on the site plan (Fig. 2.3).

The public maternity wards at FPRRH have an average of 800 deliveries per month with a c-section rate of around 20% (160/month). The adjoining post-natal and gynae wards have 40 beds. It is common to have between five and ten floor cases at any one time. There are six nursing officers and five midwives staffing the wards on three shifts with two on duty at any time. Three Senior doctors are employed to work on maternity as a whole: two medical officers and two intern doctors.

THE INTERVENTION

The intervention can best be described as an exploratory 'journey' navigated by a multi-disciplinary co-working team. K4C employs Ugandan health workers to support its activities in health facilities; with the same strict guidelines applied to professional volunteers, to ensure co-present working and guard against labour substitution. Two K4C midwives were already working in labour ward. At the start of the project, we relocated these two midwives to post-natal and gynae (PNG). They were joined initially by a UK junior doctor and several months later, a nurse from the NHS with specialist experience in wound care.[21] This team began to establish close relationships with staff on the wards and across the hospital to understand the context and observe and discuss the management of sepsis. This process stimulated an immediate focus on Infection-Prevention-Control as a preliminary to all other activity. The journey is traced in more detail in the following chapters.

Within the quite prescriptive framework of the CwPAMS funding, the priorities of the hospital and the learning gained from initial scoping work, the team re-defined its objective as follows:

> How can we improve antimicrobial stewardship in a Ugandan public referral hospital in a way that improves patient outcomes (in this case associated with maternal sepsis) and demonstrates sustainability through cost effectiveness?

METHODS USED IN THE MATERNAL SEPSIS INTERVENTION

Action research moves away from being *on* people (as objects) to being research that is participatory, *with* people and *for* people (Reason, 1988 as cited by Meyer 1993). Such an approach aligns with THET's Principles of Partnership. We noted that action research embraces the need for agency from all participants (Meyer 2000). This inclusive approach encouraged and supported individuals to make their own unique contribution to the change process, which in turn promoted group cohesion and the development of relationships between local health care workers and with overseas colleagues. Furthermore, using action research allowed us to move away from the didactic approach to learning that is commonplace in LMIC settings. In action research, the intervention moves together with the research in an iterative and reflexive process. McCormack (2015) describes this as an example of action-research 'cycles' with an action triggering a phase of research which then leads to the next action and so forth (Fig. 2.4).

---

[21] Part of the iterative quality of interventions involving deployment of UK volunteers is contingent upon recruitment processes; the characteristics and timing of volunteer stays inevitably shapes interventions. The lack of lead time in this project, as is usual, restricts the ability to plan in advance. Despite major efforts, we were unable to recruit a long-term pharmacy volunteer for the first 12 months and then the volunteer was unable to travel out due to COVID-19.

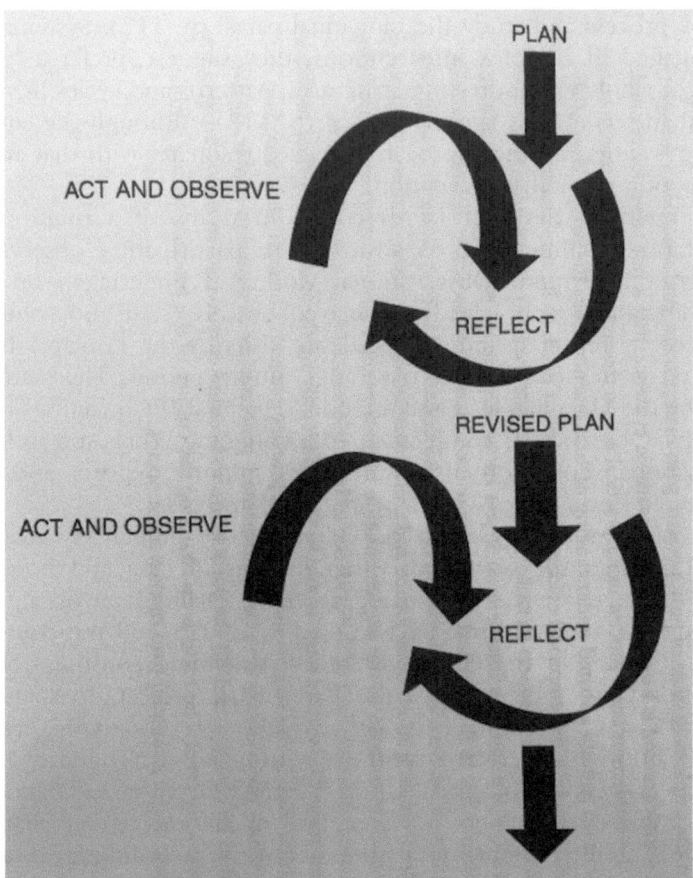

Fig. 2.4 An example of action-research cycles (*Source* McCormack [2015: 303])

Whilst this is a very helpful way of understanding action research, the idea of defined cycles with a clear starting point and closure is an over-simplification of the messiness of research (Hantrais 2009).

Mutale et al. (2016) emphasise the value of systems thinking in complex interventions in LMICs with specific attention to the generation of unintended consequences, interactions and interdependencies. They critique what they term reductionist approaches that attempt to 'dissect a complex process and study the individual parts' (p. 112). Systems theorists engaged in complex interventions, they suggest, prefer a 'general science of wholeness addressing structures, patterns and cycles in systems rather than seeing only specific events' (p. 112). Although the language they use is different, the approach proposed resonates with that adopted in this action research intervention.

Our methods can best be described in terms of a multi-method ethnography, commencing, as always, with participating observational work on the ground. Observational work was undertaken on a co-researching basis with a lead role played by K4C staff and volunteers, supported by repeated and extended site visits by the Principal Investigator and virtual co-presence over a 15-month period. The team were joined by the Ugandan lead and attended Hospital IPC meetings on two occasions. Observations complemented by ongoing WhatsApp and Skype conversations were recorded in notebooks, minutes, reports, and emails and entered into NVivo12[22] for storage and analysis.

This observational research generated theory inductively which, in turn, stimulated the search for other sources of data and honed the focus. Although we had anticipated accessing facility data on antibiotic consumption, we could not have known or understood the complexity of this process and the challenges of even defining consumption in a public hospital setting prior to the start of the project. In such situations and given the essentially inductive quality of ethnographic research, simple a priori (deductive) hypothesis setting is inappropriate. In that respect, a process of conceptualisation, theory generation and data collection took place simultaneously. Every attempt to record or collate data stimulated intense ongoing discussions about the recording processes and the nuances of its interpretation. In most cases, it led us to new lines of enquiry (theories) and approaches to data collection. Much of the data, as is normal in this context, was not collated and had to be manually and painstakingly searched for from casefiles or records books. Files were often missing or incomplete. The very poor quality of documentation in patient files and subsequent records management is a critical dimension of context with implications for AMR. Allegranzi et al.'s systematic review of health-care-associated infections in LMICS notes the lack of data and poor quality of many studies contributing to what they term the 'hidden

---

[22] NVivo is a qualitative data analysis computer software package produced by QSR International.

and serious burden on health systems and patients' (2011: 236). Study quality in their sample was related to very poor-quality record keeping and documentation as a result of 'inaccuracy of information from patients' records, and a paucity of electronic records or databases for surveillance of health-care-associated infection' (p. 235).

Data collection became a process of exploration, involving forms of local capacity-building along the way on methods of organising and storing hospital records and entering them into excel spreadsheets. In this context (as in many others), much of the facility-based data could not be interpreted at face value as facts; but rather artefacts reflecting their (social) construction. Facility data has been collected from a wide range of sources. Firstly, data on drug orders and supplies from National Medical Stores (NMS), was obtained through an online national pharmacy database, known as the Rx system, the use of which was functionalised through the project. This was supplemented by data from paper-based records (the Dispensing Log) of supplies distributed from the central hospital stores to the wards. Further, the hospital laboratory, itself supported by the Infectious Diseases Institute (IDI), proved a key partner both in the intervention itself, with laboratory results providing a critical stimulus to multi-disciplinary team working, but also in generating research data. This commenced prior to the project as part of Ackers-Johnson's microbiology doctorate (Ackers-Johnson 2020) and has continued throughout, generating valuable data on resistance patterns. The laboratory provided data on test results of samples taken from the PNG wards in 2019.

Whilst we would contend that facility-based data at FPRRH cannot be understood as 'facts' but rather social constructs contributing to a partial truth, this is not the case with the microbiology (surveillance) data generated under stringent laboratory controls. Although human error can affect the accuracy of this form of data; it is not relational in the same way as facility data. The objective status of the microbiology test results has played a powerful role in breaching disciplinary hierarchies and promoting effective team-working (see below).

This complemented a data set generated from 142 cases of suspected sepsis between January 2019 and February 2020 that were identified through a manual search of paper-based patient records. In January 2020, a phase of qualitative interviewing took place to capture perceptions of the impact and effectiveness of the intervention. Twenty-five interviews were conducted with all cadres involved in the MSI, including 50% of the nurses, midwives, intern doctors, laboratory technicians and pharmacists working on the PNG wards, two hospital managers and three UK volunteers. The interviews were transcribed and thematically analysed using NVivo 12. Ethical approval for the work was gained from the University of Salford, Makerere University, and the Ugandan National Council.[23]

Somekh suggests that not only is the action research process a continual one, it never naturally ends until a decision is taken to take stock and publish its outcomes 'to date' (2006: 6). Once again this has clear resonance with the emphasis on continuity that we feel is a feature of the principles of partnership in international development research. Although work has and will continue, this book represents the situation at the end of the initial funding period.

The following chapter presents key outcomes arising from the Maternal Sepsis Intervention and the methods outlined above.

REFERENCES

Ackers-Johnson, G. (2020). *Comparing the antimicrobial diversity of Staphylococcus aureus strains isolated from clinical cases of infection and those found as a commensal organism in Fort Portal, Uganda and further investigating the potential mechanisms of resistance present* (PhD research on-going).

Ackers, H. L., & Ackers-Johnson, J. (2016). *Mobile professional voluntarism and international development: Killing me softly?* Palgrave PIVOT. http://link.spr inger.com/book/10.1057%2F978-1-137-55833-6.

Allegranzi, B., Nejad, S. B., Combescure, C., Graafmans, W., Donaldson, L., & Pittet, D. (2011). Burden of endemic health-care-associated infection in developing countries: Systematic review and meta-analysis. *Lancet, 377,* 228–241.

Bates, P. (2014). *Context is everything.* Perspectives on Context. London: Health Foundation.

Cox, J. A., Vlieghe, E., Mendelson, M., Wertheim, H., Ndegwa, L., Villegas, M. V., et al. (2017, November 1). Antibiotic stewardship in low- and middle-income countries: The same but different? *Clinical Microbiology and Infection,* 812–818. https://doi.org/10.1016/j.cmi.2017.07.010.

Denyer Willis, L., & Chandler, C. (2018). Anthropology's contribution to AMR control. *Investment and Society,* 104–108. http://resistancecontrol.info/wp-content/uploads/2018/05/104-08-chandler.pdf.

[23] HS249ES.

Denyer Willis, L., & Chandler, C. (2019). Quick fix for care, productivity, hygiene and inequality: Reframing the entrenched problem of antibiotic overuse. *BMJ Global Health, 4*(4), e001590. https://doi.org/10.1136/bmjgh-2019-001590.

Gould, S. J. (1981). *The mismeasure of man*. New York: Norton.

Hantrais, L. (2009). *International comparative research: Theory, methods and practice*. London: Palgrave Macmillan.

Harding, S. (1991). *Whose science? Whose knowledge? Thinking from women's lives*. Ithaca, NY: Cornell University.

Harding, S. (2015). *Objectivity and diversity. Another logic of scientific research*. London: University of Chicago Press.

McCormack, B. (2015). Action research for the implementation of complex interventions. In D. A. Richards & I. R. Hallberg (Eds.), *Complex interventions in health: An overview of research methods* (pp. 300–311). London: Routledge.

Meyer, J. (1993). New paradigm research in practice: The trials and tribulations of action research. *Journal of Advanced Nursing, 18*, 1066–1072.

Meyer, J. (2000). Using qualitative methods in health-related action research. *British Medical Journal, 320*, 178–181.

Michie, S., van Stralen, M. M., & West, R. (2011). The behaviour change wheel: A new method for characterising and designing behaviour change interventions. *Implementation Science, 6*, 42. https://doi.org/10.1186/1748-5908-6-42.

Ministry of Health, Uganda. (2019, September). *The National Annual Maternal and Perinatal Death Surveillance and Response (MPDSR) Report FY 2018/2019*. Ministry of Health Uganda.

Moyo, D. (2010). *Dead aid: Why aid is not working and how there is another way for Africa*. London: Penguin.

Mutale, W., Balabanova, D., Chintu, N., Mwanamwenge, M. T., & Ayles, H. (2016). Application of system thinking concepts in health system strengthening in low-income settings: A proposed conceptual framework for the evaluation of a complex health system intervention: The case of the BHOMA intervention in Zambia. *Journal of Evaluation in Clinical Practice, 22*(1), 112–121. https://doi.org/10.1111/jep.12160.

Ngonzi, J., Tornes, Y. F., Mukasa, P. K., Salongo, W., Kabakyenga, J., Sezalio, M., et al. (2016). Puerperal Sepsis, the leading cause of maternal deaths at a Tertiary University Teaching Hospital in Uganda. *BMC Pregnancy and Childbirth, 16*, 207.

Rajkotia, Y. (2018). Beware of the success cartel: A plea for rational progress in global health. *BMJ Global Health, 3*(6), e001197.

Somekh, B. (2006). *Action research: A methodology for change and development*. Maidenhead: Open University Press.

Storeng, K. T., & Palmer, J. (2019). When ethics and politics collide in donor-funded global health research. *Lancet, 394,* 184–186.

Vogel, I. (2012). *Review of the use of 'Theory of Change' in international development.* Review Report Department for International Development https://assets.publishing.service.gov.uk/media/57a08a5ded91 5d3cfd00071a/DFID_ToC_Review_VogelV7.pdf.

Wright-Mills, C. (1959). *The sociological imagination.* New York: Oxford University Press.

CHAPTER 3

The Impact of the Maternal Sepsis Intervention

Abstract This chapter presents data on maternal mortality in Uganda and the contribution that sepsis makes to mortality. Against this backdrop, it identifies key outcomes of the intervention including major improvements in maternal mortality and reductions in the length of patient stays, readmission rates and hospital expenditure.

Keywords Antimicrobial resistance · Antimicrobial stewardship · Infection Prevention Control · Maternal mortality · Maternal sepsis

It is common in ODA projects to be required to produce 'outcome indicators' related specifically to 'beneficiaries'. As noted in the discussion on methods, the expectation is usually for quantitative metrics. The funding call and associated guidance, unusually for UKAid projects, did not specify patient outcome indicators but focused more on stewardship processes and antimicrobial consumption data. It is also more usual, in research, to discuss the intervention processes in more detail and report the findings at the end. The project proved to be more impactful than we had anticipated with significant reported impacts on maternal mortality, patient stays and hospital budgets.[1] We have chosen to report these outcomes first and

---

[1] It may be interesting to note that these outcomes were not predicted or described in the MSI application or present in the logframe as we felt they would have seemed

© The Author(s) 2020   37
L. Ackers et al., *Anti-Microbial Resistance in Global Perspective*,
https://doi.org/10.1007/978-3-030-62662-4_3

then to try to unpack the processes contributing to these outcomes. As we noted in the methods chapter, funding bodies and researchers are as much concerned with attribution (and change processes) as they are outcomes. And in such complex interventions many variables will intervene to influence attribution and outcomes. In action research, some of these variables will come directly from iterative responses to ethnographic understandings of context; others will come from external, often hidden, processes.

Maternal Deaths from Sepsis at Fort Portal Regional Referral Hospital (FPRRH)

Allegranzi et al.'s systematic review (2011) reported the very poor quality of facility data on health-care-associated infections and associated impacts:

> Unfortunately, in studies retrieved findings related to increased length of stay and attributable mortality and costs associated with health-care-associated infection were fragmentary. (2011: 236)

Qualitative interviews with health workers and stakeholders repeatedly referred to the fact that there had been no deaths from sepsis at FPRRH since the MSI became operational. Many of the respondents alluded directly to reductions in maternal deaths and the effects of this on the reputation of the wards:

> FPRRH was identified once again as the RRH in Uganda with the highest maternal mortality rate. It was in the newspapers and the Director is very concerned and embarrassed about this so any improvements on the ward are very welcome. (midwife)

Health workers were keen to cite examples of patients whose lives had been saved because of (their) improved practice. They were not only delighted and highly motivated by their ability to save these women's lives, but the experience also provided direct affirmation of the value of behavioural changes on the wards:

overly ambitious had we been required to deliver on them. This process could lead to an under-statement of impacts in ODA projects.

We have had 4 complicated cases (this week). We thought the patients would die and they walked out happy. There was a patient we thought would pass away, but she survived as she got the appropriate antibiotics. (midwife)

These responses, supported by our observations on the wards, encouraged us to undertake a review of maternal deaths. In practice, it proved difficult to find all the cases and to identify a clear causal factor in the files. After detailed analysis of the documentation, Fig. 3.1 documents a reduction in deaths from sepsis over time since implementation began in June 2019. We tracked the data back to 2018 to assess trends and this revealed some unexplained peaks and troughs.

Although most responding health workers referred to the sharp decline in deaths, analysis of the maternal mortality data indicated the persistence of deaths associated with sepsis. We could describe the method here as a forensic ethnography, and an example of where a mismatch in perceptions and data led us to delve deeper. The data support the contention that, since implementation, there have been no deaths from sepsis amongst those women having c-sections in FPRRH. One of the 'training' mechanisms the project team have instigated involves regular audit meetings on the wards; these include all cases of maternal death or serious sepsis and provide an opportunity to assess practice and monitor outcomes. Each of the sepsis deaths discussed in these meetings have involved women referred into the hospital from health facilities some distance away

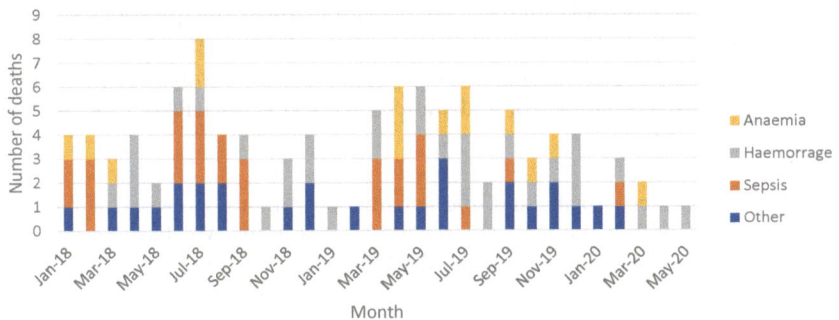

Fig. 3.1 Causes of maternal death (FPRRH) (*Source* Maternal Mortality Registers and Files, January 2018–May 2020)

primarily in refugee camps. These women face major referral pathway delays and arrive at the hospital late at night often in a very critical condition.

'THE ROAD TO DEATH'[2]: UNDERSTANDING THE IMPACT OF GEOGRAPHY AND DELAYS

Regional Referral Hospitals have a particular role to play in the Ugandan public health system. That role is predicated on the efficacy of the referral system as a whole and the functioning of lower level referral facilities so that only those cases that need treatment at a specialised facility should be referred there. This system is barely functioning. One of the major problems facing FPRRH is congestion caused by systems malfunction and significant delays in referrals with many women dying shortly after arrival. The only deaths from maternal sepsis during the project have involved late referrals from distant facilities and refugee camps. These cases either stemmed from or bypassed facilities with responsibility for Emergency Obstetric Care that should have been able to treat them in a timely and effective fashion. At the request of the hospital director, we undertook an analysis of the residence of women who die at FPRRH. This illustrates the distance many had to travel before reaching FPRRH.

Aware of the impacts of such deaths on the reputation of the hospital the Director encouraged the team to explore this issue in more detail. Analysis of a further 1000 admissions onto maternity evidenced similar referral patterns.

This data was not planned for in the application but provides critical context to an understanding of sepsis (and AMS) at the hospital. The interviews with health workers and stakeholders illustrate the problems associated with travelling long distances to access FPRRH. One intern doctor commented:

> Our ward is very small, and the volume of patients is overwhelming because the nearby hospitals are not doing c-sections. We have patients now coming from Kyenjojo hospital where they are not doing sections; they just put on the form that they have no anaesthetists.

---

[2] Filippi et al. (2005).

Other respondents talked about the problems of having women referred-in especially from Kyenjojo (50 km away) and Mubende (145 km away); facilities which have theatres and post-natal wards.

> But why are the doctors not working there? We have heard it is because the anaesthetists are not on the ground. They are bumping the cases down here simply because their anaesthetists are not there. In the referring centres you will find zero deaths – they will keep a mother there until she is now risky and between leaving there and arriving here... most women die within 1 h of arriving. (stakeholder)

Another (stakeholder) respondent referred the importance of distinguishing those sepsis cases arising from local operations and those coming in through referrals or, as is often the case, self-referrals:

> Of the patients operated in Buhinga fewer are having sepsis. We should distinguish these from those coming from outside of Buhinga; we have 2 separate entry points.

Sadly, a sepsis-related death occurred in February 2020. The case illustrates the challenges facing a hospital when women face extensive delays. In this case, the woman travelled some 77 km from Rwamwanja Health Centre, near Kamwenge. This area supports a refugee camp. The case was reported to us as follows:

> The 18-year-old woman was referred to FPRRH 6 days after an emergency C-section apparently for obstructed labour and a still birth. When she arrived late at night doctors removed a retained swab and started antibiotics. The following morning, she was found to have intra-abdominal sepsis and needed another operation to wash the pus out and remove her uterus which was the source of the infection. You could smell the infection from metres away. We took her to theatre, washed out her abdomen, and removed her uterus, which was severely infected and necrotic. We then put her on meropenem, our strongest possible antibiotic. She later died.
>   Causes of death: Poor monitoring of labour and delayed caesarean at the health centre, leading to foetal death and increased risk of infection. Likely poor asepsis and retained swab during the caesarean leading to post-operative infection. Delay in recognition that she was not getting better resulted in delays in referral to FPRRH. Possible antibiotic resistance meant that any initial empirical treatment was not enough and possibly there was

even resistance to meropenem (though she might have been too far gone by the time she reached us).

Figure 3.2 shows that six of the maternal deaths in 2019 came from this refugee camp as well as a large volume of admissions (Fig. 3.3). Respondents have advised us that, although health facilities in such camps are typically better resourced and ring-fenced UNHCR funds are available for transport to FPRRH, drivers obtain greater funding for night visits and are therefore inclined to delay transportation of very sick patients until night-time. Other respondents suggested drivers may wait until they have 3 or 4 cases to transport, especially if any of those cases are local Ugandans who do not come with a budget. It has also been reported that where local Ugandan women do manage to get access to this transport to FPRRH they are not permitted to travel back via this route. Whatever the reasons, late referrals such as this one leave little opportunity for staff at FPRRH to save lives and this also directly impacts antimicrobial stewardship.

The impact of late referrals on maternal mortality from sepsis is confirmed by Ngonzi et al. (2016) in Uganda and Kongnyuy et al. (2009) in Malawi. Seni et al. (2013) put another twist on this relationship between late referrals and SSI incidence reporting an 'alarming' prevalence of SSI in Uganda amongst 'emergency surgeries which accounted

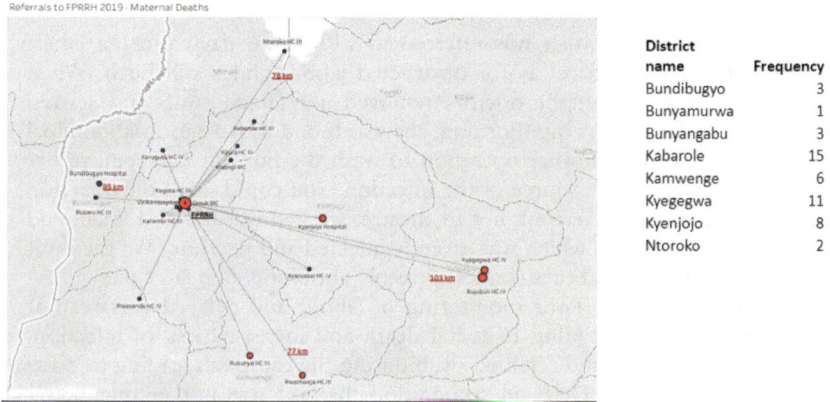

District name	Frequency
Bundibugyo	3
Bunyamurwa	1
Bunyangabu	3
Kabarole	15
Kamwenge	6
Kyegegwa	11
Kyenjojo	8
Ntoroko	2

Fig. 3.2 Distances travelled to FPRRH (maternal mortality cases) (*Source* Analysis of maternal mortality cases, 2019)

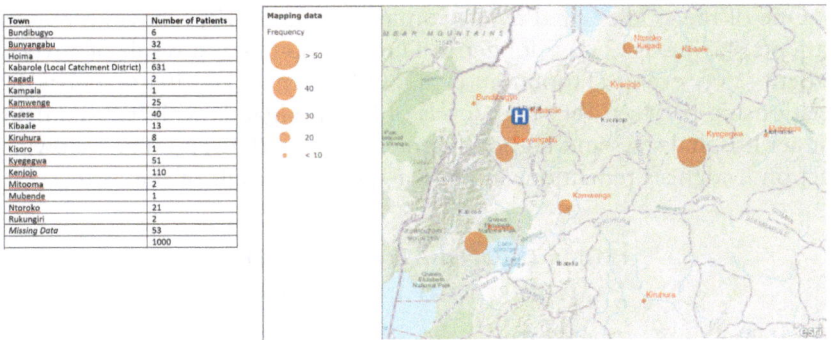

Town	Number of Patients
Bundibugyo	6
Bunyangabu	32
Hoima	1
Kabarole (Local Catchment District)	631
Kagadi	2
Kampala	1
Kamwenge	25
Kasese	40
Kibaale	13
Kiruhura	8
Kisoro	1
Kyegegwa	51
Kenjojo	110
Mitooma	2
Mubende	1
Ntoroko	21
Rukungiri	2
Missing Data	53
	1000

Fig. 3.3 Distances travelled to FPRRH (Admissions, 2019) (*Source* Analysis of 1000 Maternal Admissions, 2019)

for more SSIs cases as opposed to elective surgeries' (2013: 5). Very few Health facilities in Ugandan run elective list for non-fee-paying cases. All the referrals coming into FPRRH are for obstetric emergencies. It is easy to understand the hospital director's concerns about these cases and the attribution of blame. The relationship between adverse patient safety events and reputational damage is noted by Slawomirski et al. (2017) in a report on the economics of patient safety (2017). The authors cite the results of several studies in High Income Settings illustrating the impact of adverse patient safety events on prolonged hospital stays. They conclude:

> Hospital-acquired sepsis accounts for a large proportion of additional hospital days, standing out as one of the most expensive and most clinically complex conditions to treat. (2017: 16)

The data presented above and the discussion about referral cases could be seen as a distraction from the main focus of the research. We believe it illustrates the importance of a very deep and intuitive approach to context. On the one hand, the ability to evidence the source of sepsis cases has supported local policy and underlined the need to invest in these referral facilities and, potentially, extend the intervention to such sites. On the other, the data and mapping process has served an important function in challenging the basis of media coverage of the hospital (and its position in maternal mortality league tables). This has proved a real motivation for hospital managers and health workers alike.

One of the studies cited by Slawomirski et al. (2017) (above) reported patients staying 10.2 days longer in Dutch hospitals as a result of adverse patient safety events (Hoonhout et al. 2009). Seni et al. (2013) report the relationship between very high levels of SSI in Uganda with 'substantially longer hospital stays and higher treatment costs' (2013: 6). The following section evidences reductions in patient stays as a result of the MSI.

Impacts on Patients, Their Families, and Livelihoods

Spending time in hospital is a significant burden in low resource settings especially amongst the poorest of patients. During their time in hospital, they are unable to undertake any economic activity or caring responsibilities themselves and often have up to five or six attendants residing in or near the hospital to support them, providing food and routine care. Many patients admitted to a Regional Referral Facility will have travelled some distance from their home village. The large number of women described in the notes as 'running away' (or self –discharged) prior to completion of their treatment[3] (and at high risk of readmission or mortality) do so in order to care for children and family and to enable their families to return to their activities. Extended stays in busy referral facilities also contribute directly to congestion. This presents a challenge in terms of maintaining adequate IPC. Congestion is often manifested in floor cases making it hard to clean facilities effectively and for over-stretched staff to deliver efficient and respectful care. The fact that every patient will have a number of attendants many of whom sleep and eat amongst the patients on the floors is a further hazard. Zulfiuqar et al. (2013) reported concerns about the security and infection risks posed by the presence of in-patient attendants at a tertiary hospital in Karachi. Most patients in their study had at least one attendant with 85% attendants staying for the duration of the patient stay and performing critical care duties for patients including feeding, washing and administering of medications. Extended stays of patients and accompanying attendants create the conditions for the spread of Healthcare Associated Infections some of which have shown multi-drug resistance (*Acinetobacter*). Where extended stays are due to delays

---

[3] These cases contribute to the poor quality of antibiotic consumption data as we have no way of knowing if they complete their prescribed dose.

in closing wounds (secondary closure),[4] this also creates opportunities for secondary infections.

Most respondents expressed the view that patients were staying for shorter periods on the ward and they were experiencing fewer readmissions due to sepsis. The following midwife suggests that the project has had a significant impact on length of stay:

> These days the patients do not stay on the ward for very long. At first those ones with sepsis they could spend 2 weeks or 4 weeks on the ward but now you find a mother has come with sepsis – you do the dressing - a mother will not even spend a week on the ward. Because they remove a swab – they take it there – we get the results – the doctor comes and investigates and the good thing we are doing closure now in our room – we are not now taking them to theatre because thanks to K4C we now have the instruments to do them there and to sterilise them.

This observation is backed up by the laboratory:

> Patient stays have really reduced on the ward and we produced evidence on this last year. Patients were staying for over 30 days on the ward prior to the project – now in many cases after culture and sensitivity testing[5] and closing the wounds they can then go home after 10 days. We have reduced patient stays on the ward and even in terms of costs, shorter stays and fewer attendants on the wards.

One midwife explains how reductions in length of stay are a combined result of improved wound care, improved prescribing in response to laboratory results and the infrastructural improvements that have improved the functionality of the evacuation room. This involved the provision of additional instruments, sterilising equipment and storage cupboards and has enabled intern doctors to close (re-stitch) wounds on the ward

---

[4] Secondary closure is a term used on the ward to describe cases where the original c-section wound opens as a result of infection. In some cases, the staff have to re-open infected wounds. Some of the wounds on the ward gape openly with a diameter of up to 10cms. They then must be debrided (removal of damaged tissue) and dressed until the infection is gone and they are able to re-close them.

[5] Culture and Sensitivity Testing is the term used to describe the process of taking a sample from a patient; growing the bacteria from that sample in a laboratory and then testing the susceptibility of (or responsiveness to) any bacterial growth to available antibiotics.

without sending women to surgical theatre. Theatre days for elective wound closure are only available twice a week and if a woman misses that date, she may have to remain on the ward with an open wound for another 4 or more days. The savings in such cases include not only shorter stays but reduced use of precious theatre time. As one doctor put it, 'wound closures cannot be a priority for theatre time'. This also reduces the risks associated with moving patients around the hospital. An intern doctor describes the changes he has witnessed on the ward:

> The instruments in the evacuation room means we can now do procedures that used to take a long time as we had to sterilise equipment between cases so mothers can be treated much more quickly, get better and leave the next day. This takes pressure off theatre too. We now dress (the wounds) and when they are ready, we do the secondary closures – they don't need to go to theatre. We have been able to move women out of the ward more quickly which has reduced congestion as women heal more quickly now. We do it in the evacuation room and now we disinfect it and we know we have to reach higher standards so we can close the wounds there. The doctors are doing most things now – the interns will do some without supervision apart from difficult wounds then they take to theatre and the senior doctors are there. They will assess the wounds and we do what we can here.
>
> [Were the doctors doing this before?]
>
> Actually, they were doing some closures, but it would take some time – they had to wait to disinfect the equipment but now it is streamlined. It has really improved now; the numbers of mothers that need to go to theatre has reduced a lot – because what happens – the mother comes with a bad wound - these guys do the dressing and manage the mothers and the wound becomes small and easy to suture so we can do it here now on the ward. The sepsis has really gone down but maybe out of 5 we can now take 2 to theatre and the other 3 we can close the wounds here. Before the number was big and all mothers were being taken to theatre to close the wounds. In fact, in the last 2 months we have had a situation where the septic mothers' section is empty – the place is clean – you have done a great job.

Another intern refers to improvements in patient recovery and length of stay:

> What [MSI] has brought to reality is that – actually when samples are taken and followed up then things get better more quickly – as observed you will

note that this side has very limited sepsis now – there are few and even those that are there are improving so quickly and we are doing secondary closure so quickly. Before we had patients on the ward with a wound for over a week and as you bring in a new patient, they get a hospital acquired infection but now hospital stay has been reduced from 2 weeks to 5 days… we really could quantify this. We now have fewer infected wounds to dress and fewer secondary closures and we can do many of these now on the ward. Before every infected patient was cross-infecting the next patient – the longer they are on the ward the more this happens. We have outgrown waiting to go to theatre.

The doctor also refers to the benefits of having a new trolley and instruments but, importantly, the impact of leadership and improved human resource management:

Most of the secondary closures are being done on the ward now. Now we have the instruments and the trolley – before there wasn't anyone who would take responsibility to wash the instruments so there was a delay – now the instruments are enough – the challenge was for someone to take responsibility for washing and sterilising them. Now we do the closures as soon as they are ready – before the room was not sterile for secondary closures but now it is ready. There are now few cases going to theatre unless it is more serious like a burst abdomen. Before in the evacuation room the instruments were lying in jik (disinfectant) so you had to wait. Then the patients can become septic again. Now you can't spend a day waiting.

Another midwife adds the observation that this has also reduced the volume of readmissions:

The wounds get better and then the interns can do the secondary closures and we have had no evidence of readmission of these cases.

When the project started, we were advised that many of the cases on the post-natal and gynae wards were re-admissions; that is women who had been treated on that ward and discharged and then had to return due to infection. In March 2020, the midwifery team decided to track every woman coming onto the ward with an infected wound for a period of 3 weeks. Of the 121 cases tracked, there was no evidence of re-admission.[6] These perceptions are backed up by the evidence drawn from facility records. In a pain staking process and with support from the hospital administrator, the team have worked with the records office to reorganise a mass of files into date order and entered data into a spread-sheet of every case of sepsis in 2019–2020. This has provided a rich database enabling us to assess, objectively, the impact of the intervention on patient stays.

IMPACT ON PATIENT STAYS

The following section reports on an analysis of all cases of *suspected* sepsis recorded over a 13-month period from January 2019 up to and including January 2020. 142 cases were identified in patient records which recorded patient stays from the date of operation (usually a c-section) to discharge. A t-test was done to determine whether there was a difference in hospital stays between those patients who had received a culture swab and those who had not. It was found that hospital stays were significantly lower in patients for whom culture swabs were completed ($M = 21.28$, $SD = 1.17$) than those for whom this was not done ($M = 27.35$, $SD = 1.67$): t $(128) = 2.99$, $p = 0.003$. On average, the group that was swabbed stayed in hospital for **6 days less**, following their operation.

The records included another potential 'start' date, the date of diagnosis. A t-test was also performed to determine whether there was a difference in hospital stays from the date of diagnosis, between those patients who had received a culture swab and those who had not. It was found that hospital days were significantly lower in patients for whom culture swabs were completed ($M = 13.44$, $SD = 1.08$) than those for whom this was not done ($M = 19.38$, $SD = 1.51$): t $(129) = 3.31$ p $= 0,.0016$. On average, the group that was swabbed stayed in hospital for **5 days less**, after suspected diagnosis of sepsis. Figure 3.4 shows the overall pattern of patient stays since the project commenced.

---

[6]We do not have statistics from the pre-intervention phase on this. Women were tracked, where possible post-discharge by telephone. Tracking proved difficult in around 20% cases with some women not responding to calls. However, re-admission would have been evident through ward-based reporting.

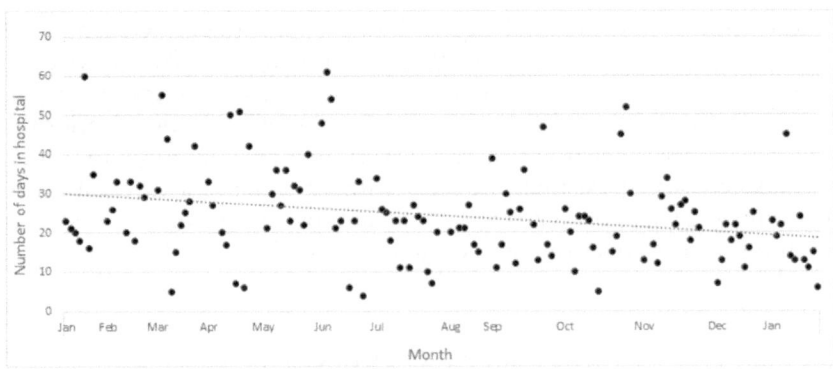

Fig. 3.4 Length of stay based on date of operation (*Source* Patient Records PNG Ward FPRRH [January 2019–January 2020])

As noted above, these reductions in patient stays will be associated with multiplier reductions in the volume of attendants staying in and around the wards.

IMPACTS ON HOSPITAL BUDGETS

Finally, and inevitably, extended stays represent a major cost to the hospital in terms of infrastructure, staffing and consumables. The effect of these improvements has been welcomed by the Senior Administrator who comments on the efficiencies this has generated:

> When sepsis is managed then the resources used to manage these patients reduce drastically. By reducing long stays which brings about savings you have contributed to the hospital budget with real term savings. That's an area for me now to look into. We should be able to see in real terms how much savings we have been able to make.

Allegranzi et al. (2018) noted the paucity of studies identifying in any systematic way, the costs involved in managing health care associated infections and the cost-savings associated with effective interventions. The MSI did not set out to undertake detailed cost-benefit analysis. Nevertheless, as the project has evolved, and cost savings were remarked upon by respondents and senior managers, we started to explore the potential for integrating cost-benefit analysis in future work. This would provide critical impetus for sustainability and future investment. Our findings confirm Weber et al.'s assertion that strengthening WASH services in Togo was cost-minimal. Although key barriers cited by Weber's respondents referred to costs this was more a matter of institutional priorities than actual costs which are in reality very low and, importantly, could be managed within existing budgets, 'without significant external financial or material support' (2018: 1). The authors also recommend greater attention to cost-effectiveness and budget expenditures in future research (p. 64).

PATIENT EXPERIENCES

It is difficult to gauge patients' perceptions of the project's impact mainly because most will have nothing to compare their experience to. However, several midwives reported positive patient feedback:

> The mothers have noticed the project because we had a mother who had sepsis. It was her second child and she had delivered in Buhinga before but that time she didn't have sepsis. She said, 'hey you people this ward has changed'. In the first pregnancy she delayed on the ward. This time she got sepsis, so we moved her from post-natal to gynae and she said 'hey you nurses why have you changed? You have changed me to gynae ward – now what is taking place?' We told her be calm and after a few days we worked on her and she went home. She said the hospital has changed! Nowadays they know.

The in-charge also noted genuine changes in the attitudes of staff towards their patients:

> Some used to treat a death of a mother as if it were the death of a chicken. Now they see our patients as their sisters or mothers and care for them. Before some staff were hard to deal with; they barked at patients. The care in this ward is critical care and a woman could lose her life in one minute. We need the best and most caring staff.

The following chapter examines some of the constituent interventions and impacts that, together, contributed to the improvements described above. We start with a discussion of infection prevention and control. The 'control' dimension and the focus on wound management merges seamlessly into a discussion about antimicrobial surveillance and stewardship.

REFERENCES

Allegranzi, B., Nejad, S. B., Combescure, C., Graafmans, W., Donaldson, L., & Pittet, D. (2011). Burden of endemic health-care-associated infection in developing countries: Systematic review and meta-analysis. *Lancet, 377,* 228–241.

Allegranzi, B., Aiken, A. M., Zeynep, K. N., Nthumba, P., Barasa, J., Okumu, G., et al. (2018, May). A multimodal infection control and patient safety intervention to reduce surgical site infections in Africa: A multicentre, before-after, cohort study. *The Lancet Infectious Diseases, 18*(5), 507–515. https://doi.org/10.1016/S1473-3099(18)30107-5.

Filippi, V., Ronsmans, C., Gohou, V., Goufodji, S., Lardi, M., Sahel, A., et al. (2005). Maternal wards or emergency obstetric rooms? Incidence of near-miss events in African hospitals. *Acta Obstet Gynecol Scand, 84,* 11–16.

Hoonhout, L., de Bruijne, M. C., Wagner, C., Zegers, M., Waaijman, R., Spreeuwenberg, P., et al. (2009). Direct medical costs of adverse events in Dutch hospitals. *BMC Health Services Research, 9,* 27. https://doi.org/10.1186/1472-6963-9-27.

Kongnyuy, E. J., Mlava, G., & van den Broek, N. (2009). Facility-based maternal death review in three districts in the central region of Malawi: An analysis of causes and characteristics of maternal deaths. *Women's Health Issues, 19*(1), 14–20.

Ngonzi, J., Tornes, Y. F., Mukasa, P. K., Salongo, W., Kabakyenga, J., Sezalio, M., et al. (2016). Puerperal Sepsis, the leading cause of maternal deaths at a Tertiary University Teaching Hospital in Uganda. *BMC Pregnancy and Childbirth, 16,* 207.

Seni, J., Najjuka, C. F., Kateete, D. P., Makobore, P., Joloba, M. L., Kajumbula, H., et al. (2013). Antimicrobial resistance in hospitalised surgical patients: A silently emerging public health concern in Uganda. *BMC Research Notes, 6,* 298.

Slawomirski, L., Auraaen, A., & Klazinga, N. (2017). *The economics of patient safety.* OECD.

Weber, N., Patrick, M., Hayter, A., Martinson, A. L., & Gelting, R. (2018). A conceptual evaluation framework for the water and sanitation for health facility improvement tool (WASH FIT). *Journal of Water, Sanitation and Hygiene for Development, 9*(2). https://doi.org/10.2166/washdev.2019.090.

Zulfiuqar, B., Salam, A., Firoz, M., Fatima, H., & Aziz, S. (2013). Effects of inflow of inpatients attendants at a tertiary care hospital—A study at civil hospital Karachi. *Journal of Pakistan Medical Association, 63*(1), 143–147.

CHAPTER 4

Infection Prevention Control (IPC) and Antimicrobial Resistance (AMR)

Abstract This chapter outlines a key component of improved AMR; namely infection prevention control (IPC). It addresses some of the issues most commonly associated with IPC including hand hygiene, waste disposal and infrastructure. It then addresses wound management as an Infection Control issue. The emergence of wound management as a central focus in the Maternal Sepsis Intervention proved pivotal in shaping the pathway to antimicrobial stewardship.

Keywords Infection Prevention Control · Maternal sepsis · Wound management · Surgical Site Infection · Health Care Acquired Infection · Hand hygiene

'PREVENTION FIRST'

In 2017, the WHO adopted a Resolution focused on improving the prevention, diagnosis, and management of sepsis. Reinhart et al. underline the importance of recognising sepsis as a global health priority. They go on to suggest that progress towards 'a world free of sepsis' requires recognition of the key role of prevention (2017: 416). Infection prevention reduces the overuse of antibiotics which drives resistance.

© The Author(s) 2020
L. Ackers et al., *Anti-Microbial Resistance in Global Perspective*,
https://doi.org/10.1007/978-3-030-62662-4_4

Infection-Prevention-Control (or 'IPC') has featured strongly on the global health agenda for many years with significant emphasis on those infections that patients (and health workers) acquire within health facilities. Whilst attempts to tackle the source of infection in the home and workplace have formed the basis of HIV-awareness and public vaccination programmes, the phenomena known as 'Health Care Acquired Infection' (HCAIs) and its cousin, 'Surgical Site Infection' (SSIs) have focussed concern on 'adverse' events associated with hospitalisation.

What Is a Health Care Acquired Infection?

The World Health Organisation's Patient Safety Fact File (2019) defines 'Health Care Associated Infection' as follows:

> Health care-associated infections, or "nosocomial" and "hospital" infections, affect patients in a hospital or other health-care facility, and are not present or incubating at the time of admission. They also include infections acquired by patients in the hospital or facility but appearing after discharge, and occupational infections among staff.[1]

The Fact File reports health care associated infection rates of 10% amongst hospitalised patients in LMICs (2019: 9). Allegranzi et al.'s systematic review of health-care-associated infection in developing countries found that the prevalence of health-care-associated infection is 'much higher' in LMICs than HICs and concluded that Surgical Site infection was the 'leading health care-associated infection in the developing world' (2011: 28).

What Is a Surgical Site Infection?

The Centre for Disease Control (CDC) defines a Surgical Site Infection (SSI) as, 'an infection that occurs after surgery in the part of the body where the surgery took place' and occurring within 30 days after the procedure (or 12 months in the case of orthopaedic implants).[2] Seni et al.'s (2013) analysis of 314 SSI cases at Uganda's national referral hospital reported a much higher incidence of SSIs amongst

[1] https://www.who.int/gpsc/country_work/gpsc_ccisc_fact_sheet_en.pdf.

[2] https://www.cdc.gov/hai/ssi/ssi.html.

women (76.1% of their sample) and a preponderance of cases in obstetrics and gynaecology wards (62.1%). Caesarean-section and laparotomy[3] accounted for more than three quarters of all surgical procedures in their study. They conclude:

> The predominance of SSIs in obstetrics and gynaecology wards is quite alarming and thus, a need to institute stringent infection prevention and control measures in this setting, more especially in emergency surgeries which accounted for more SSIs cases as opposed to elective surgeries. (2013: 5)

A similar study in Tanzania (Mawalla et al. 2011) reported SSI rates of 26% amongst patients undergoing surgery and similarly noted the gendered impact of SSIs largely reflecting the volume of women having surgery in the first place.[4]

Following the award for the MSI, the funding bodies added an additional 'request' that each team conduct a Global Point Prevalence Survey (G-PPS).[5] This is a standardised survey of antimicrobial use amongst all in-patients in a hospital on a given day designed to deliver comparative data for international benchmarking. It captures that data through documentation in patient records (which may be an inaccurate reflection of practice). The MSI team undertook the GPPS on May 7th, 2019. It involved all 42 patients on the post-natal and gynae wards at 8 am (22 in gynae and 20 in post-natal). The GPPS found that 94% patients on post-natal ward were prescribed antibiotics for surgical prophylaxis (94%) with one case involving suspected Community Acquired Infection (CAI).[6] The picture in the adjoining gynae ward was quite different. Here, 45%

---

[3] A laparotomy is a surgical incision into the abdominal cavity, for diagnosis or in preparation for major surgery and is commonly used in obstetrics and gynaecology cases.

[4] Mawalla recorded a significantly higher SSI rate amongst those men who did have surgery which they suggested may be due to multiple risk factors such as smoking and HIV infection.

[5] http://www.global-pps.com/ The value of the GPPS process is discussed below.

[6] Patients who come into the hospital with a pre-existing infection are regarded as having acquired that from the 'Community' which may imply their home environment. In many cases, these patients will have acquired an infection from a previous health facility and referred on to the hospital. In such cases CAI and HCAI are conflated and difficult to distinguish.

Table 4.1 Antibiotic Prescribing on Post-natal and Gynaecology Wards in FPRRH (GPPS)

	Gynaecology	Post-natal
Total no of patients	22	20
Percentage of patients on antibiotics	45	90
Percentage of antibiotics for Community acquired infection	20	6
Percentage of antibiotics for Health-care-associated infection	50	0
Percentage of antibiotics for medical prophylaxis	0	0
Percentage of antibiotics for surgical prophylaxis	10	94
Percentage of antibiotics for unknown indication	20	0

Source Results of G-PPS, May 2019 as reported to FPRRH IPC Committee

patients were prescribed antibiotics and 50% of these were related to a suspected health care associated Infection (Table 4.1).

INFECTION PREVENTION AND CONTROL IN THE UGANDAN NATIONAL ACTION PLAN

The Ugandan National Action Plan on AMR outlines its 'One Health' approach arguing that, 'Prevention is the most effective, affordable way to reduce risk for and severity of resistant infections' (2019: 6). Strategic Objective Two identifies key actions to improve infection prevention and control. These span four inter-linked areas: IPC in healthcare facilities; IPC in the community; biosecurity in agriculture and vaccination programmes. Action 3.2.1 sets out key objectives to 'Strengthen Infection Prevention and Control Programs in Healthcare Facilities' and is the area of most direct relevance to the Maternal Sepsis Intervention (Fig. 4.1).

Objective 3.2.1 reiterates well-rehearsed (if neglected) IPC goals: to raise awareness; improve hand hygiene, basic infrastructure, and waste disposal. Goals 4 and 6 add a specific AMR 'twist' and illustrate the immediate connection with the surveillance objectives outlined in Strategic Objective 4. The creation of guidelines to limit the spread of multidrug-resistant organisms and timely diagnosis and treatment of drug-resistant organisms requires strong multi-disciplinary team working with laboratory scientists, pharmacists, doctors, nurses and midwives.

This chapter reports first on the more familiar aspects of IPC concerned primarily with creating an environment on the post-natal and

1. Maintain up-to-date infection prevention guidelines and standards of professional practice and ensure their availability in all healthcare facilities
2. Institute/strengthen and support minimum standards for infrastructure in healthcare facilities that promote IPC
3. Institute/strengthen and support proper functioning of Infection Prevention Control (IPC) and Medicine Therapeutic (MTC) committees
4. Create and promote specific guidelines for limiting the spread of multidrug-resistant organisms
5. Support availability and proper use of infection prevention materials and supplies
6. Encourage timely diagnosis and treatment of drug-resistant microorganisms
7. Promote hand hygiene and other hygienic practices and behaviours that prevent transmission of infectious disease.
8. Promote campaigns for infection control at healthcare facilities
9. Institute systems of incentives or rewards that monitor and uphold good IPC practices
10. Promote safe waste disposal and safe treatment practices in healthcare facilities
11. Create and strengthen coordinating activities at all levels from local level facilities to the Ministry of Health for IPC
12. Improve human resource systems, education, and commitment to professionalism

Fig. 4.1 Objective 3.2.1 Strengthen Infection Prevention and Control Programmes (*Source* Ugandan National Action Plan on AMR)

gynaecology wards that reduces opportunities for women entering that ward to acquire a health care-acquired infection as a direct result of practices on that ward. The 'control' component of IPC is often neglected; IPC is not just about prevention; it is also about controlling infection. Many women arriving on the PNG will already have been exposed to risks of HCAI and SSI either in the operating theatres and labour wards at the same hospital or in the referring facilities they pass through on their journey into the hospital. In such cases, the focus on the PNG is on early identification of infection and appropriate management. Wound management has emerged as a key concern in the control of infection for those women with infected wounds; for the women and attendants around them and for the health workers caring for them as they become a source of infection to others.

IPC in the Maternal Sepsis Intervention[7]

The project team were aware of the central importance of IPC to antimicrobial resistance when we applied for funding. Arguably the emphasis in the literature on AMR has focused too much on the management of

[7] A version of this chapter has been written up as a policy document for the hospital and includes photographs of the ward and more details of the various interventions and associated costs. This can be found at www.knowledge4change.org.uk/.

antibiotics by individual health workers and patients which, to use a colloquial expression from the UK, amounts to, 'locking the door after the horse has bolted'. Denyer-Willis and Chandler emphasise the importance of preventive approaches:

> Antibiotics have become …. a quick fix for hygiene in settings of minimised resources. (2019: 1)

This is particularly relevant in LMICs, where, they argue, antimicrobials are, '*put to work to correct the fractured infrastructures of care, water and sewage, hygiene and demands for ever increasing [health worker] productivity*' (p. 2).

We have seen this in previous K4C work on antibiotic stewardship. Women at a health centre III were being routinely prescribed prophylactic antibiotics following vaginal birth as a mechanism to protect against uncertainty surrounding hygiene and sanitation in both the hospital setting and the home setting (Welsh 2019).

Denyer-Willis and Chandler (2019) emphasise the importance of 'connectivities' and underline the need for multi-disciplinary teams and methods (including social science and anthropological approaches) in order to present an holistic and accurate picture of the deeply contextual factors contributing to AMR and potential responses. Prevention must be the starting point of all holistic AMR interventions; it is also the most cost-effective.

HAND HYGIENE

Maina et al. describe **wa**ter, **s**anitation, and **h**ygiene (WASH) as the key foundations of AMR in Kenyan hospitals:

> Poor WASH increases hospital-associated infections and contributes to the rise of antimicrobial resistance. (2019: 1)

The survey tool developed by Maina et al. and piloted in 14 general hospitals in Kenya showed major performance variations between hospitals and wards reflecting differences in the built environment, resource availability and leadership. They identify waste management and (healthworker) hand hygiene as 'critical indicators' with hand hygiene achieving an aggregate score across all facilities of only 35% (p. 1). Allegranzi et al. report even lower levels of hand hygiene compliance of around 20% in LMICs (2011: 235). Hand hygiene compliance at FPRRH was observed to sit at a mere 17.4% in 2018 (Mbabazi 2018).

K4C previously held a grant (awarded in 2015) from the Tropical Health and Education Trust which focused on IPC in the Kabarole region. In keeping with our experience of behaviour change in Uganda, the hand hygiene project combined formal training with continuous mentoring whilst commencing quality controlled local manufacture of alcohol-based hand sanitiser. Providing training without ensuring that health workers had access to the opportunity to exercise that knowledge was, we felt, arrogant and insulting. The hand hygiene project achieved a considerable shift in health worker behaviour; but only for as long as K4C was in a position to fund the costs of the hand gel. Despite persuasion, facilities proved unwilling to contribute in any way at all to the constituent (and cheap) ingredients for hand gel production. When the MSI commenced, we immediately expected and noticed the absence of hand gel. The juxtaposition of a broken and empty hand gel dispenser next to the Ugandan guidelines on hand hygiene on entry to the ward can only have had a demotivating impact on health worker behaviour especially when dealing with highly infectious patients (Fig. 4.2).

Denyer Willis and Chandler echo this sentiment:

> The saddling of responsibility for hygiene with individuals who have limited ability to change the environment in which 'good hygiene behaviour' is expected to operate leaves these individuals to find solutions that are more feasible and within their control, such as the use of antibiotics. (2019: 3)

Another potential solution to the individual risk the health workers face is to decide not to uncover and dress infected wounds (see below). The COM-B behaviour change model presented in the Application Guidance (Fig. 1.1) emphasises the importance of 'opportunity' to behaviour change. In our experience providing training in hand hygiene without access to resources fails to translate into improved behaviour (unsurprisingly) and acts as further demotivation as it underlines the failure of institutions to honour their duty of care to protect employees.

This chapter reports on the process of improving the IPC infrastructure. This has been achieved through continual discussion and co-decision-making. Informed by our previous experiences of formal training (Ackers et al. 2016), the team resisted the temptation to wade

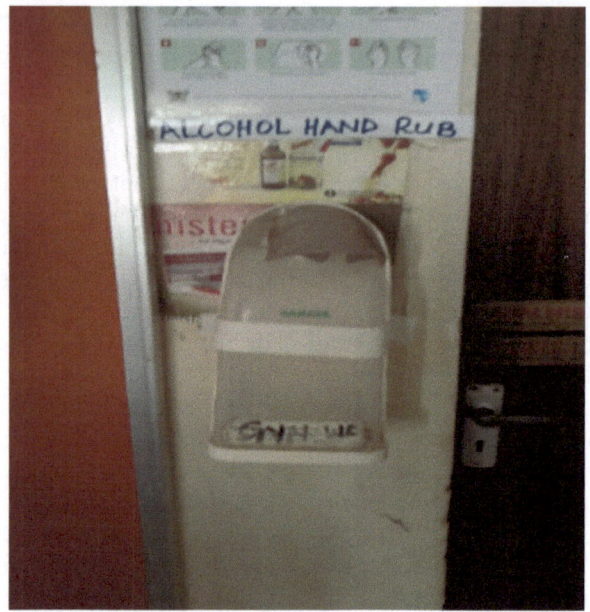

Fig. 4.2 Empty and damaged hand gel dispenser next to Ugandan guidelines on handwashing

in and 'train'. This does not mean that no education or knowledge creation/mobilisation took place. Rather that it evolved through team working.[8]

Concerns about hand hygiene and the sustainability of our (previous) intervention stimulated a proposal (at the start of the project) to the hospital which would have supported the co-production and co-financing of IPC consumables and infrastructure repairs to support implementation of the hand hygiene protocol throughout the whole hospital. The proposal was based on principles of Public–Private Partnership, as envisioned by the Ministry of Health's Strategic Plan. Unfortunately, at that point, the hospital felt unable to agree to this proposal.[9] However, given

---

[8] Our approach to knowledge mobilisation in the MSI is discussed in more detail in Chapter 8.

[9] We return to the PPP discussion in Chapter 9.

the existence of project funding and the importance attached to reducing infection risks both for patients and health workers, the team agreed to continue providing hand gel for the ward. This has involved placing more robust dispensers on the walls in key locations (such as the sepsis area) but also, in December 2019, providing health workers with their own refill-able personal dispensers. One of the intern doctors interviewed on the ward showed us his dispenser, attached to his uniform. It is interesting to see how he specifically refers to using the gel after a procedure and to protect himself:

> IPC is improving - even when we cannot wash our hands, we all now have our own hand gel so after any procedure or examination we use this alcohol. It has made life safer for us. What they send through National Medical Stores is not enough – at least now we have this. This has really helped us improve patient care.

The mechanism used to improve hand hygiene on the ward has priori-tised infrastructural repairs and supplies of running water, soap and hand gel. We firmly believe that Ugandan health workers are aware of the importance of hand washing both to their own well-being and that of their patients. Formal training courses (referred to locally as Contin-uing Medical Education or CMEs) conveying that knowledge at this stage in the project would have been inappropriate. We are also aware that health workers in Uganda (as in the UK) do not always exercise exemplary behaviour (they do not apply their knowledge to practice). In our experience, effective and continuous role modelling and mentoring in the context of good leadership is the only way to build an IPC culture. The project team included a Ugandan medical educationalist and midwife who had previously worked with us on the hand hygiene project and undergone high-level training in IPC through the Infection Control Africa Network (ICAN) programme.[10] This midwife is Ugandan and comes from the local region so is fluent in the main local language (Rutoro). This has enabled her to build excellent relationships on the ward supporting staff with wound dressing and other duties whilst also developing a contextually appropriate version of the WHO Hand Hygiene

---

[10] This involved a one-week training program in South Africa and another in Cameroon with an emphasis on infection prevention to improve on hand hygiene, sanitation and health care waste.

Compliance Tool and Infrastructure Audit Tool.[11] The in-charge nurse noted the effectiveness of this approach and, perhaps surprisingly, the fact that local staff did not feel threatened by her presence:

> [K4C midwife] is on the wards at times watching them hand wash and pulling them up. Not criticising but making them constantly aware of the importance of hand washing to themselves and the patients.

One of the intern doctors noticed the improvement:

> The project has really improved on IPC. These days it is a must to clean your hands and staff are using the hand sanitiser.

The following midwife echoes a concern we were familiar with from our previous project; namely the challenge of drying wet hands where there are no disposable towels[12]:

> We have improved hand hygiene because we have the sinks repaired and we have enough alcohol sanitiser – there is soap and running water. Now if you don't wash your hands that is your attitude.
> [Do people wash their hands now then?]
> Actually, it has changed – some do, and some feel hand washing takes time to dry but with the sanitiser you can move quickly between patients and wash hands after the procedures. There are no towels, so we use pieces of gauze.

The K4C midwife suggests that whilst much improvement has been made on hand hygiene compliance there is room for more:

> I'm seeing a bit of improvement – they are using hand sanitiser – I can identify this from the WHO forms. I can identify areas they tend to forget. At least most of them remember the hand sanitiser but there is gap when they go to a new patient. Those who forget – previously they worked on

---

[11] https://www.who.int/gpsc/5may/moment1/en/.

[12] The Hand Hygiene project piloted the use of re-usable single-use hand towels. This has remained effective only in the smaller Health Centre 3 facility that K4C is partnering with. In FPRRH, the risks associated with abandoned dirty towels and the demands of apportioning staff to wash them in the absence of laundry machines outweighed the potential gains.

patients without having it in mind to clean their hands and some of them it is still in their trait.

[So, should we put more sanitisers on the walls?]

It is working for some, but we are looking at the distance between the beds and the sanitiser and we need one fixing to the wall in the sepsis area.

The WHO Hand Hygiene Observation Tool[13] was used to audit hand hygiene compliance. Observation took place during day shifts where most procedures are performed. The common procedures include operations such as bed-cleaning, patient examination, wound dressing and drug administration. Hand Hygiene compliance was assessed twice; first in October 2019 and secondly, in March 2020 (Table 4.2).

Table 4.2 Hand Hygiene Compliance in October 2019 on PNG ward at FPRRH

Cadre	Hand Washing			Hand Gel		
	Opportunities	Actual	Compliance (%)	Opportunities	Actual	Compliance (%)
Midwives	54	18	33	54	23	43
Intern Doctors	36	2	6	36	8	22
Senior Doctors	12	2	16	12	4	33

Source Adapted WHO Hand Hygiene Compliance Audit

The results show relatively poor adherence to WHO Hand Hygiene targets in the first observation period. We did not assess compliance prior to the project so this will represent a marked improvement on the previous period especially when sinks were not working, and hand gel was not present on the wards. It is interesting to note that compliance with hand gel use is stronger than hand washing, and midwives and nurses have higher compliance rates than doctors. Maina et al.'s study in Kenyan hospitals reported qualitative findings suggesting that, 'nurses are more conversant with infection prevention issues' and complaints by nurses that, 'doctors don't embrace the issues of IPC' (2019: 13).

[13] https://www.who.int/gpsc/5may/tools/en/.

Table 4.3   Hand Hygiene Compliance in March 2020 on PNG ward at FPRRH

Cadre	Hand washing			Hand sanitiser		
	Opportunities	Actual	Compliance (%)	Opportunities	Actual	Compliance (%)
Midwives	26	20	76	26	23	88
Intern Doctors	12	7	58	12	11	91
Nurses	12	8	66	12	7	58

Source Adapted WHO Hand Hygiene Compliance Audit

Table 4.3 shows the results for a second phase of observation in March 2020. By this point, all staff had personal hand gel bottles as well as access to dispensers at the entrance to every ward. The observer in this phase distinguished nurses and midwives (which we had not done previously) and as is common, no senior doctors were present on the ward during observations.

The nurses and midwives received informal feedback after the previous observations and were aware that observation had been taking place. Table 4.3 shows marked improvement in the second observation phase especially amongst intern doctors who show a strong preference for using alcohol gel with compliance increasing from 22 to 91%, and hand washing, from 6 to 58%. Higher levels of compliance were observed among midwives than nurses. The placement of hand sanitiser on the trolleys, on walls and in the health workers' possession have improved compliance. Observation also indicated very strong compliance with guidelines on the use of gloves (which are generally available). In both observation periods, glove utilisation was at 100%. Table 4.4 identifies compliance rates at key opportunities for hand hygiene in the second observation.

Analysis of the 56 'moments' identified during this observation period showed that 95% of health workers washed hands *after* body fluid exposure and 75% remembered to wash their hands *after* touching the patient's surrounding. Some chose to use hand gel rather than hand washing (Fig. 4.3).

Figure 4.2 gives a flavour of how the tool has worked in practice. It also shows differences in the use of hand washing and hand gel. Before touching a patient, there is a mix of staff using hand washing and hand

Table 4.4 Hand Hygiene 'Moments' and Compliance (March 2020) on PNG ward at FPRRH

Action/'Moment'	Number	Percentage
Before approaching a new patient	09	45
Before aseptic action with patient	13	65
After body fluid	19	95
After completing a procedure with a patient	15	75
Total	56	

Source Adapted WHO Hand Hygiene Compliance Audit

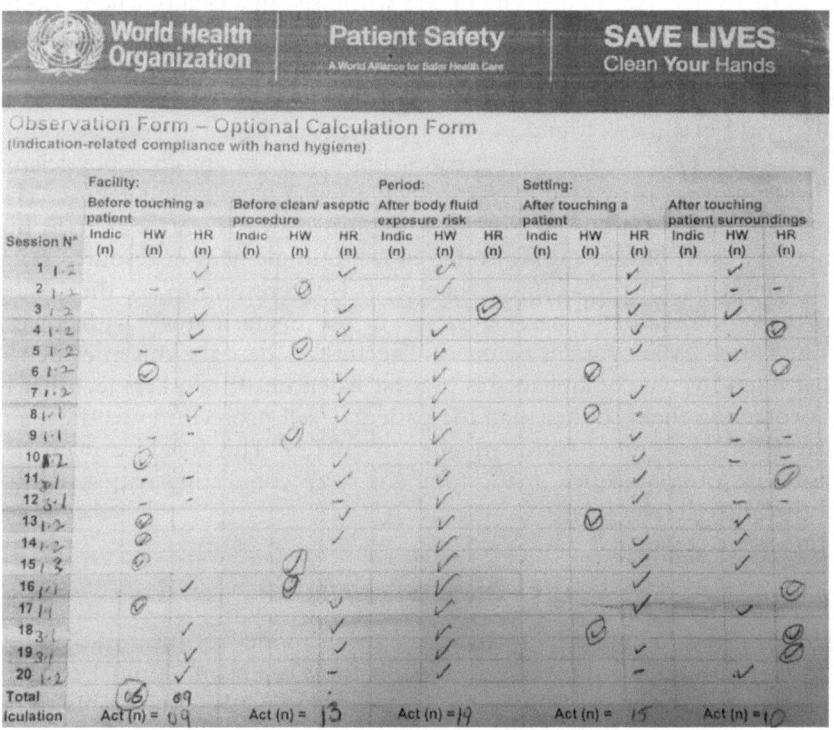

Fig. 4.3 An example of the WHO Calculation Form (March 2020 Observations)

gel whereas staff are more likely to use hand gel before commencing an aseptic procedure. Nearly all staff washed their hands after exposure to body fluids. Hand gel is much more likely to be used after touching a patient or touching patient surroundings. The results indicate a high level of compliance with IPC advice and efficient combination of the two dimensions of hand hygiene. Monistrol et al. identify hand hygiene as the 'most important procedure' in preventing HCAI (2011: 1212). Their intervention in a tertiary hospital in Spain, focused mainly on education and training, demonstrated marked improvements in hand hygiene compliance (also using the WHO Tool) with compliance improving from 54.3 to 75.8%. This improvement was witnessed in a facility where rooms were shared by only 2 or 3 patients and alcohol hand gel was available at every bedside. Given the far more restricted access to hand gel in the PNG, the rate of improvement is remarkable. It is interesting to note that Monistrol et al. also found poor physician compliance in comparison to nurses and, post-intervention greatest improvement amongst physicians. These findings are echoed in our study particularly with hand gel use. The use of hand gel on unsoiled hands is described by Monistrol et al. as the 'new standard of care'; it is also far easier to implement in resource-poor environments with weak infrastructure. It is interesting to see the results of the '5 moments of hand hygiene' in the Spanish study with lowest compliance '*before* patient contact'. The authors describe this as evidence that hand hygiene compliance is highest when health workers feel that it is protecting them (rather than the patient); 'self-protection was the main driver for performing hand hygiene' (p. 1217). This would explain the high rate of compliance in the moment *after* connecting with patients' body fluids (Table 4.4).

IPC INFRASTRUCTURE

The emphasis on identifying an intervention model for future scalability led to continuous improvement of audit tools to ensure optimal contextualisation. This meant that results are not directly comparable. The action-orientation of the MSI and our concern to develop and trial appropriate tools was more important than achieving a controlled, comparative, sample. We are also very aware that K4C presence on the wards may have contributed a 'Hawthorn Effect'[14] (as reported by Monistrol et al. 2011). Our concern to identify optimal methods of knowledge mobilisation, through co-presence and co-working relationships no doubt accentuates this effect. The presentation of hand hygiene audit data here is to illustrate trends; the quantitative data are heavily influenced by the intervention and we firmly believe that this type of data is best complemented by on-going observation and qualitative interviewing. Clearly the infrastructural investment made in supplying hand gel was a major driver of change. The provision of hand gel was a 'quick fix' preliminary intervention that stimulated an unfolding identification of other infrastructural IPC concerns.

The WASH FIT initiative is one of the World Health Organisation's responses to critical concerns about Patient Safety.[15] Weber et al. (2018) identify key aspects of infrastructure in LMICs that undermine progress in improving patient safety. They cite a study by Cronk and Bartram (2018) based on aggregated information from 78 LMICs which reports that 50% of health facilities studied lacked piped water; 33% did not have improved sanitation; 39% did not have soap and water for hand washing and 39% lacked proper medical waste management.

During preliminary observational phases of the MSI, we were increasingly aware of serious infrastructural challenges that would undermine our ability to bring about those behaviour changes required to reduce and manage hospital acquired infections effectively. The project team, led by the midwives on the ground, made an initial assessment of the ward infrastructure which was later followed up during a field visit by the project leads. This confirmed the need for certain investments to promote IPC on the ward, in the designated sepsis area and in a partitioned area (known as the evacuation or procedure room) used for minor procedures such as wound closing which we later identified as an intervention focal point. The team also discussed the contextual relevance of the existing WHO Infrastructural Survey Tool[16] which colleagues felt was too focused on

[14] The Hawthorne effect is a term used to describe the possibility that individuals involved in an experiment or study modify their behaviour in response to their awareness of being observed.

[15] The WHO define Patient Safety as 'the absence of preventable harm to a patient during the process of health care', https://www.who.int/patientsafety/en/.

[16] https://www.who.int/gpsc/5may/tools/evaluation_feedback/en/.

hand hygiene and neglected issues like signage and the status of furnishings (closing cupboards and doors etc.). We also had concerns about the section in the WHO Tool on provision of clean drinking water. This is very rarely provided in Ugandan public health facilities unless a local donor does so. We were acutely aware of the value of providing access to clean drinking water to ensure women are hydrated. However, having discussed this with local staff and the ward in-charge we made the decision not to provide clean drinking water given the serious problems associated with large numbers of visitors on the wards (see below). One of the K4C midwives reported back to the UK lead on this on-going discussion:

> I talked to [the in-charge] about the issue of drinking water for patients. She did not welcome it, based on the behaviours of both patients and attendants. She felt the attendants are unruly, and she looked at the possibility of them using the drinking water for brushing and other things. She also felt sustainability may be a question. The other concern was IPC related and we should not commit ourselves by introducing drinking water on the ward.

This illustrates the ethnographic quality of the work continuing even at a distance and the value of this approach in guarding against the unintended consequences (externality effects) of seemingly easy and benevolent interventions. Some 12 months later, stimulated by ongoing concerns about the impacts of poor hydration on wound healing in several severe sepsis cases, we agreed to provide a large water boiler in the nursing station to enable staff to boil tap water and provide it to patients. This intervention may obviate the need for cannulation in cases where intra-venous fluids are given in Uganda as a substitute for oral hydration but carry their own HCAI risks (and costs).

We also had concerns about the sections of the WHO Tool auditing provision of paper towels and a waste basket for used paper towels. To our knowledge, no public health facilities in Uganda have paper towels. We therefore decided to remove these two sections of the WHO Tool from our audit. The Modified Infrastructure Audit Tool includes a list of key components with a scoring column that supports a quantitative overall score for audit and comparison purposes.[17]

---

[17] The team extended use of this tool across the whole hospital to support of COVID-19 intervention.

IMPROVEMENTS IN THE INFRASTRUCTURE SCORE

A first audit, using the original WHO Tool, was completed in October 2019. This identified concerns around the sterility of equipment and instruments, ready access to hand sanitiser and healthcare waste. We also noted the absence of adequate hand-washing facilities, lack of labelling of soap bottles and display of hand hygiene posters. By this time, the project had already provided hand gel. The general patient environment was also much improved. However, we noted weaknesses under the equipment heading in terms of broken trolleys, cleaning of medical devices and re-use of single use items. Although gloves were generally available and used, there were concerns about access to eye shields and protective masks. Concerns were also expressed about the quality of waste management with health workers failing to sort waste properly and waste bins often over-loaded with poor adherence to management of sharps. These observational findings echoed Maina et al.'s conclusion that waste management is one of the weaker aspects of IPC (2019). Another continuing and persistent challenge has been in improving very poor-quality documentation and record-keeping. The Infrastructure Audit was repeated 3 times; in October 2019, January 2020 and March 2020. The overall results are presented in Table 4.5.

Audit 3 raised several residual issues (inadequate cleaning of equipment between patients; lack of labelling of waste boxes and lack of data on in-service IPC training). The following section summarises the 'problems' identified. Interventions took place in cycles gradually reducing gaps and improving the opportunities for effective and holistic IPC. Quite often as one problem is solved another emerges either simply because it opens a new process but also because of the kinds of externality effects referred to above.

The intervention kicked off with an early decision to re-decorate the ward. This was largely cosmetic but gave the opportunity for a thorough

Table 4.5   Results of IPC Infrastructure Audit

	Score	Percentage
Audit 1 (October 2019)	37/63	58.7
Audit 2 (January 2020)	50/63	79.4
Audit 3 (March 2020)	61/63	96.8

Source Infrastructure Audit

deep clean. The process was also designed as a motivational exercise to promote staff and patients' sense of well-being and team working. In response to concerns about the 'openness' of the existing sepsis area wooden doors were fitted to isolate the sepsis area and try to prevent attendants sleeping and eating on the floor. The state of the mattresses on the ward was a great concern to heath workers as the condition enabled body fluids to enter and soak into exposed foam. The team purchased 20 new waterproof mattresses and an additional 12 waterproof covers for the mattresses that were least damaged. In response to a request from midwives and nurses, hard wearing washable aprons were provided for staff to use during wound dressing procedures to protect themselves and the patients.

Chapter 6 discusses the problem of stock-outs in the hospital not only of drugs but also essential IPC consumables. At regular intervals, the hospital runs out of 'JIK', a bleach used ubiquitously in differing dilutions for many aspects of cleaning and disinfection in Ugandan hospitals. As a rule, K4C does not provide consumables to facilities as we believe this to be unsustainable. However, we did make a personal donation of 2 large bottles when the hospital ran out. One of the problems in the use of JIK was not simply its absence, but the tendency to use it far too concentrated, which damages materials and equipment. Use of JIK in this way had caused major damage to examination beds, hospital screens and instruments on the ward. This illustrates the merits of continuous observational engagement and an acute understanding of context. When we asked the hospital pharmacist about this, he reported that health workers found it hard to understand the formulae for JIC concentration on the wall in PNG (below). When we assessed this guideline, we immediately understood the major weakness in this attempt at science communication expressed as a knowledge gap on the part of nursing staff. The guideline was replaced with a simple plastic measuring jug, marked to guide the proportion of JIK to water, and buckets and training in the use of these and the problem was immediately resolved (Fig. 4.4).

Continuous engagement on the ground has enabled us to identify some of these simple problems and rectify them. The issue of dilution was picked up again once the project had purchased a new examination couch. K4C has witnessed very rapid rusting on such beds in the past where staff use neat JIK to clean metalwork. Pharmacy have provided advice on appropriate dilution of JIK for cleaning of infrastructure and furnishings and advised cleaning the bed with hand gel. Advise was also

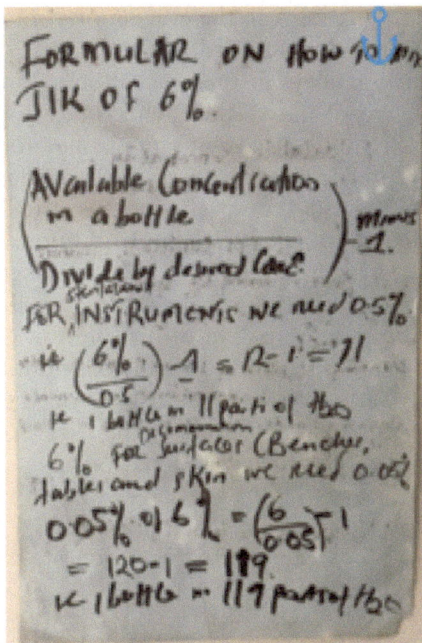
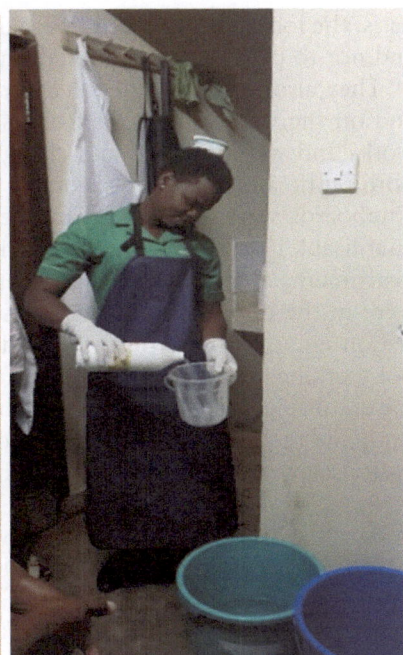

Fig. 4.4 Formulae for JIC dilution

provided on the correct dilution of hydrogen peroxide which is used when dressing septic wounds. When not correctly diluted, this can cause harm by delaying healing processes. Staff are now continually trained and mentored using this guidance.

REPAIR OF SINKS

Staff were conscious that there were only 2 sinks on the wards and one of these, in the evacuation room where secondary closures and other procedures take place, was not functioning. Lengthy discussion took place about the pros and cons of providing an additional sink. Although this immediately seemed like a good idea to the foreign team, local staff recognised that sinks can themselves become a source of infection, especially when simultaneously used as a sluice for disposal of body fluids. On that basis, the local team decided to repair the one sink in the evacuation room and not to provide an additional sink on the main ward.

They also identified the need for new trolleys in the evacuation room and on the ward as existing trolleys failed to move (their wheels did not work) and were very old and rusted. Staff also used the same trolley for both septic and non-septic wounds (a practice which has now changed). Cupboards were also placed in the evacuation room and stainless steel (sterilisable) galley pots were provided to maintain sterility. Additional instruments were purchased so that work could continue when one set were in the process of being sterilised. At that time, there was no sterilisation equipment on the ward and the only autoclave in maternity was not working. Midwives had to walk up to the surgical ward and wait until the sterilising unit there was available wasting valuable staff time on the ward and reducing the use of the procedure room. The following excerpts indicate appreciation of interventions and the impact on behaviour:

> Nowadays I'm happy because we have sterile swabs. We didn't have instruments, so you put your hands in (the wounds). Now we have instruments we can do it properly.
>
> [Do you feel that protects you as well?]
>
> Yes, because nowadays we have really improved on hand washing. We have hand gel, aprons and masks. They are giving hand gel now to improve on infection because after dressing you use gel and you are free of infections. At first, we didn't have those things.
>
> [When you said some of the midwives didn't want to change dressings because they were smelly, do you think they were also worried about catching infections themselves?]
>
> Yes, because at first when you said, 'who is going for dressing?" she would reply; 'Who is going there? I am not protected.' They were fearing infection from the patient to them. You would go there without an apron (gloves were there) but still most people would not like to go there. Now this has improved (Midwife).

A laboratory scientist echoes this view:

> IPC has changed for the best. This is very positive and has been sustained through support with materials.

The following midwife notes benefits in terms of health worker safety and productivity:

The new trolleys have really helped. We can now sort our equipment out and this eases our work. The instruments in the evacuation room means we can now do procedures that used to take a long time to do as we had to sterilise equipment between cases. Mothers can be treated much more quickly, get better and leave the next day. This takes pressure off theatre too. We now have enough instruments and receivers to use (a receiver is like a kidney dish). We had very few so if we used them, we had to wait for another cycle to sterilise them before we used them again.

K4C staff expressed concern that the system of disinfecting instruments requires further improvement:

> In the evacuation room some of the instruments are being left lying in jik – they should only be left in jkc for 10 minutes. We are trying to improve on it. They know they should remove the instruments from the jkc. If they put them in, by the time they have finished getting the patients ready to go back on the ward they should take them out. We had that discussion in the last staff meeting and it was agreed that if the doctors have finished the procedure, they have to make sure they alert the nurses to wash the instruments and hand over the evacuation room and then they should lock it.

This quote shows how IPC issues are raised at most (every) staff meeting, not always as a distinct training intervention but within the normal course of events. At this time, midwives had to carry instruments, gauze, etc. to surgical theatre for sterilisation. This involved staff leaving the ward and often waiting in surgical theatre for the steriliser to be available. The project has now provided a dedicated autoclave to speed this process and reduce unnecessary movement of people and instruments between wards. The discussion above illustrates the incremental and progressive approach we have taken, gradually identifying and responding to many small issues to improve both IPC and productivity. Despite the above interventions, the in-charge remained concerned at the level of over-crowding:

> We are currently faced with 'overwhelming numbers' and yesterday there were 56 patients in a ward with a bed capacity of 40 so this is an on-going problem. This meant that there were many floor cases in the gynae area and into the sepsis zone.

This overcrowding is continuing to occur despite reduced patient stays. In many respects, it is outside of the hospital's control as many patients continue to be referred from other health facilities including Hospitals and Health Centre IV facilities who should be able to cope with c-sections and post-natal patients. Overcrowding, combined with lack of hospital beds leading to floor cases, and inadequate sterilisation of hospital tools are factors that Denyer-Willis and Chandler (2019: 3) suggest contribute to HCAI and unnecessary prophylactic antimicrobial prescribing.

INFECTION PREVENTION AND CONTROL AND WOUND MANAGEMENT

Flexibility and reflexivity are critical to high impact action-research. These are respected qualities in complex intervention research (Moore et al. 2015; Richards and Hallberg 2015). As noted above, this can cause some creative tension with funding bodies who, for accountability purposes, are keen to adhere as closely as possible to projected activities, associated budgets, and international protocols. The success of the MSI has derived from its very grounded, inductive, approach and the acute attentiveness to contextual dynamics. The comments of the Senior Administrator that, '*No little thing is ignored*' capture this attention to context and to processes on the ground that together contribute to antimicrobial resistance and shape our ability to respond effectively to it.

Although the decision to focus on surgical site infections and take samples for laboratory testing had been planned for some time, the rationale for this focus was that many of these women would be otherwise well and that this would enable us to identify hospital acquired infections. And, although we had planned to swab c-section wounds as the basis for the laboratory testing and our AMR surveillance activity, we had not anticipated how important wound care itself was to infection control and the management of antimicrobial resistance. The opportunity to engage more actively in wound management, as a key constituent of an holistic AMR intervention, was neither mentioned in the Call Specification, in our application or indeed in the National Action Plan. It came about as a result of the recruitment of one of the first UK volunteers to the project; a nurse who had extensive experience of working on surgical site infection studies in a London hospital and a strong interest in wound management. This midwife was a 'diaspora' volunteer; a Ugandan national fluent in the local language (Rutoro). This, coupled with her commitment K4C's approach to active co-working, really helped to establish rapport. On arrival, she immediately noticed patients with very badly infected, gaping wounds, and poor practices in terms of wound care. As in the case of IPC, rather than immediately commence 'training' or design protocols, she worked alongside local staff to understand the context within which SSI wounds were developing and contributing to sepsis. The team spent an intensive two weeks engaging in structured observation and follow-up of 71 women who had had a c-section at FPRRH during that period. This enabled them to identify a number of concerns including; the lack of any consistent approach to the cleaning, dressing and swabbing of wounds; patient observations (essential to the early identification and management of infection and sepsis) and prophylactic antibiotic use (with prophylactic doses not completed in 94% of observed cases).

This initial observational phase also identified concerns about IPC processes (discussed above) including the use of unsterilized gauze for wound dressing; the disposal of infectious waste; the lack of sealed containers and cupboards for storage of sterilised gauze and instruments and poor management of materials on badly rusted and dysfunctional trolleys. Analysis of the 71 cases revealed a re-admission rate of 10% confirming the findings of subsequent interview data that many women were leaving the ward and returning several days later with badly infected wounds.

It was for this reason that the team embarked on intense, continuous, mentoring on wound care. One of the local midwives had already taken an active interest in wound management which she described as the most neglected area on the ward. She suggested that, when she arrived on the ward, her focus on wounds was perceived negatively:

I found the ward stinking. There was so much sepsis I went to where the smell was worst. The staff were running away from the bad smell. Some women stayed for over 2 months. There were staff shortages and many patients. I said, 'let me look at these wounds' so I started there. Some midwives were dressing wounds, but they were reluctant. The work was too much. It depends on someone's interest, but it was a major priority for me. Gynae was somehow neglected; everyone was shying away because of the smell. They knew it was smelling but didn't know what to do.

Staff were encouraged to get into the habit of documenting and reading patient notes, dressing wounds, and using simple tools developed to observe and evaluate the effectiveness of their approach. They created their own medication records following receipt of laboratory results and started a Ward Report Book. The in-charge echoes the observation discussed above:

> The staff are now identifying and dressing wounds; when I arrived and before the project started the midwives didn't do this. This resulted in a terrible smell throughout the ward which has now gone; you can smell the place is better?

She commented on the work K4C had introduced on wound cleaning:

> Before that the midwives didn't do it – midwives often focus on the pelvis[18] and not on bedside nursing. I am a midwife and a nurse and appointed as a nurse in my role. Midwives would have been taught the theory of dressing wounds in their training but had never used those skills in practice. K4C staff really encouraged staff to start to identify wounds and treat them. The number of cases going back to theatre as a result of infected wounds has dropped significantly; they can now be better managed on the wards. This has been important in decongesting theatre and was better for the mothers. All the local staff are now engaged in wound dressing; there has been a real change in staff attitudes. Before there was no one to remind them of their knowledge and skills. They did not have the idea to manage wounds, some did not have the knowledge and there was resistance by midwives who felt it wasn't their role. Then those of [K4C] came and this has made the job a lot more pleasant; people are enjoying work more.

It is wonderful to see not only the impact on patients' wounds and the decongestion of the theatre but also to hear that health workers on the ward were beginning to enjoy work; this is the environment that creates meaningful opportunities for behaviour change. Another local midwife who has become actively involved in wound dressing speaks of how practice in this area has been transformed and suggests that this is also a reflection of improved IPC and provision of basic materials which protects them:

---

[18]We assume that the reference to pelvis here is not anatomical as such but an indication that midwives focus on deliveries rather than wound management.

Wound dressing has totally changed. At first, we used not to dress the wounds every day for sure. Nurses didn't want to dress the wounds because they were stinking. Sepsis had increased on the wards but nowadays staff want to do dressing because the wounds are not smelling like at first.

[Did you know how to do wound dressing?]

Yes, we were taught how to sterilise gauze and do dressings but the problem when we reach the ward you just stop because you don't have things to use on the wards. Nowadays I'm happy because we have sterile swabs. We didn't have instruments so you put your hands – now we have instruments – we can do it properly. At first when they said 'who is going for dressing' she would reply; 'I am not protected so they were fearing infection from the patient to them so you go there without an apron – gloves were there – but still most people would not like to go there. Now this has improved because we now do dressings twice a day, they don't get smelly.

Another local midwife, also actively involved in wound dressing on the ward, was proud to report on the case of a mother who 'ran away' from another health centre (Kamwenga, a 65 km distance); 'She came here and we dressed her wound and were able to make her better'. She said that she used to dress wounds but had learnt new techniques and the importance of documentation and, when a patient appears unwell, of taking vital signs[19]:

We just tried to clean the wounds. With the guidance of the K4C colleagues we really now know that we have to take vital observations. We call them baseline bedside observations. Then if the temperature changes or the pulse we know something is happening, so we know to do vitals. We document them to find if they are stable or not. With the help of [K4C midwife] we have time. When these people are there, we can do this so right now we do. I knew wound dressing before, but I have learnt higher techniques and also recording exactly what you see on that day in the notes. Documentation is sometimes a problem, but we are trying to improve. If you come and dress the wound and don't document no one else will know but now it helps team working. Your colleagues will also come and if you REALLY document then someone else will come – and if the wound is still bad after 3 days of wound dressing we can ask – why is this wound not getting better?

[19] Whilst in other settings vital signs will be taken for all patients, the practice in this context is to take vital signs when a patient is showing signs of deterioration.

One of the intern doctors also remarked on improved wound dressing comparing practices favourably to other hospitals he had worked in:

> Wound management has improved greatly. Before and in other hospitals I've worked in we recommend twice a day dressing and they are not changed even once. Wound management has improved greatly, and mothers are receiving twice daily dressings, so they improve so quickly. The staff have developed a **trait to inherit** – it's a great impact.

The use of the phrase 'trait to inherit' would indicate a degree of continuous behaviour change or culture change in practices on the ward. The point about the smell on the ward is repeated by many respondents both in interviews and in casual conversations on the ward. Certainly, the ward no longer smells, and this has improved the working environment for all staff and patients. There has been lengthy debate on the ward about the use of honey and sugar in wound care. Whilst the use of honey is more widely accepted as having an evidence base (and antimicrobial qualities) the use of sugar, instigated by one midwife following several years of exposure to wound treatment in other settings, has received conflicting views. The midwives on the ward are clear that applying both sugar and honey to the wounds speeds healing and significantly reduces odour. Intern doctors and pharmacists are less convinced of sugar's healing properties with one referring to its use as 'bush medicine'. Having said that there was no suggestion that using sugar had a negative effect.[20]

This chapter has addressed the issue of infection-prevention and the contribution that quite simple and cost-effective interventions can have in reducing the incidence of infection on the wards and managing those infections that do exist more effectively. Improved IPC reduces the volume of wound infections per se. This is clear from the marked reduction in readmissions onto the ward with infected wounds. Wound management is also a dimension of IPC with an emphasis on the control aspect. And it is this attention to observing and managing wounds that created the opportunity for collection and analysis of antimicrobial resistance and, subsequently, antimicrobial use.

---

[20] This is an ongoing debate which we will return to in due course.

REFERENCES

Ackers, H. L., Ioannou, E., & Ackers-Johnson, J. (2016). The impact of delays on maternal and neonatal outcomes in Ugandan public health facilities: The role of absenteeism. *Health Policy and Planning*, 1–10.

Allegranzi, B., Nejad, S. B., Combescure, C., Graafmans, W., Donaldson, L., & Pittet, D. (2011). Burden of endemic health-care-associated infection in developing countries: systematic review and meta-analysis. *Lancet, 377*, 228–241.

Cronk, R., & Bartram, J. (2018, April). Environmental conditions in health care facilities in low- and middle-income countries: Coverage and Inequalities. *International Journal of Hygiene and Environmental Health, 221*(3), 409–422.

Denyer Willis, L., & Chandler, C. (2019). Quick fix for care, productivity, hygiene and inequality; reframing the entrenched problem of antibiotic overuse. *BMJ Global Health, 4*(4), e001590. https://doi.org/10.1136/bmjgh-2019-001590.

Maina, M., Tosas-Auguet, O., McKnight, J., Zosi, M., Kimemia, G., Mwaniki, P., et al. (2019). Evaluating the foundations that help avert antimicrobial resistance; Performance of essential water sanitation and hygiene functions in hospitals and requirements for action in Kenya. *PLoS ONE, 14*(10), e0222922. https://doi.org/10.1371/journal.pone.0222922.

Mawalla, B., Mshana, S. E., Chalya, P. L., Imirzalioglu, C., & Mahalu, W. (2011). Predictors of surgical site infections among patients undergoing major surgery at Bugando Medical Centre in Northwestern Tanzania. *BMC Surgery, 11*, 21.

Mbabazi, W. (2018). *Assessing compliance with infection prevention and control guidelines among various cadres of healthcare workers at a regional referral hospital in Uganda* (Unpublished MSc thesis).

Monistrol, O., Calbo, E., Riera, M., Nicolas, C., Font, R., Freixas, N., et al. (2011). Impact of a hand hygiene educational programme on hospital-acquired infections in medical wards. *Clinical Microbiology and Infection, 18*, 1212–1218.

Moore, F. G., Audrey, S., Barker, M., Bond, L., Bonell, C., Hardeman, W., et al. (2015) Process evaluation of complex interventions: Medical Research Council Guidance. *BMJ, 350*.

Reinhart, K., Damiles, R., Kisson, N., Machado, F. R., Schachter, R. D., & Finfer, S. (2017). Recognizing sepsis as a global health priority — A WHO resolution. *New England Journal of Medicine, 377*, 414–417.

Richards, D. A., & Hallberg, I. R. (2015). *Complex interventions in health*. London: Routledge.

Seni, J., Najjuka, C. F., Kateete, D. P., Makobore, P., Joloba, M. L., Kajumbula, H., et al. (2013). Antimicrobial resistance in hospitalised surgical patients: A silently emerging public health concern in Uganda. *BMC Research Notes, 6*, 298.

Weber, N., Patrick, M., Hayter, A., Martinson, A. L., & Gelting, R. (2018). A conceptual evaluation framework for the water and sanitation for health facility improvement tool (WASH FIT). *Journal of Water, Sanitation and Hygiene for Development, 9*(2). https://doi.org/10.2166/washdev.2019.090.

Welsh, J. (2019). *Providing an evidence base for antibiotic stewardship for midwives in the Kabarole District of Uganda: A modified action research study* (Unpublished PhD thesis).

World Health Organisation. (2019). *Patient safety fact file.*

CHAPTER 5

The Role of Microbiology Testing in Controlling Infection and Promoting Antimicrobial Stewardship

Abstract This chapter discusses the Role of Microbiology Testing in Controlling Infection and Promoting Antimicrobial Stewardship. It details the role that culture and sensitivity testing has played in creating the 'objective' evidence base that facilitated and nurtured midwifery empowerment, task shifting and multi-disciplinary team working. The chapter presents findings of resistance patterns of isolates from women with suspected sepsis.

Keywords Antibiotic resistance · Culture and sensitivity testing · Isolates · Rational prescribing · Empirical therapy · Antibiogram

Objective 3.2.1 of the Uganda National Action Plan (Fig. 4.1) outlined key components of IPC. All the familiar 'good housekeeping' aspects of IPC are listed. Sub-Objective 6 encouraging 'timely diagnosis and treatment of drug-resistant microorganisms' stands out as a less usual element of IPC. It expresses the 'control' dimension of IPC and forms an essential interface with the stewardship aspects of the National Action

---

Some of this material has been published in Ackers et al. (2020).

© The Author(s) 2020
L. Ackers et al., *Anti-Microbial Resistance in Global Perspective*,
https://doi.org/10.1007/978-3-030-62662-4_5

Plan demonstrating the need for holistic approaches to AMR. Objective 3.3 of the Action Plan combines a concern to 'Optimize Access to Effective Antimicrobial Medicines' with access to 'Diagnostics in Human Health'. This latter objective overlaps with Sub-Objective 6 (above) and is focused on (1) ensuring availability of affordable and accurate diagnostic tools to all health facilities and (2) enhancing systems for financing access to diagnostics and antimicrobial medicines. Achieving progress in Sub-Objective 6 requires identification of the microorganisms present in infected wounds using appropriate diagnostic tools and using those results to make evidence-based and contextually relevant clinical and prescribing decisions. The success of this rests fundamentally on high-functioning, multi-disciplinary teams.

We noted (above) how the Call for Funding identified 3 of the 10 Fleming Fund objectives as priorities. This explicitly excluded the Fleming Fund Objective of 'building laboratory capacity for diagnosis'. Furthermore, the reference to multi-disciplinarity did not refer to laboratory science. However, the outcomes expected from projects did include 'the use of microbiology data to inform decision-making'. This illustrates both the mixed messaging coming from the Call and the underlying challenge of attempting to isolate specific components of AMR.

Most public hospitals in Uganda lack the capacity (equipment and consumables), human resource and expertise to undertake effective, timely and free microbiology testing. One of the intern doctors interviewed had trained and worked in Mbale Regional Referral Hospital in north-east Uganda (520 km from Fort Portal). This enabled him to draw interesting comparisons between the two facilities in swabbing practices:

> The main thing we are doing here is blood culture, swabs and culture and sensitivity testing so at least we are able to prescribe the best antibiotics.
> [Was it like that in Mbale?]
> No – it wasn't easy. We did culture and sensitivity testing there but we had to use a private laboratory that charged 30,000 shillings per case[1]. About 50% of people were able and willing to pay for this service. The hospital microbiology laboratory did not function; they lacked consumables, culture medium ….
> [Did patients understand the reason for the testing?]
> Yes, they did – it was more a question of money.

[1] About £6.

FPRRH is an exception. The laboratory at FPRRH is supported by the Infectious Disease Institute (IDI) with funding from a range of international partners. As such, it offers excellent laboratory facilities. The IDI team has invested considerably in encouraging health workers in FPRRH to send samples to the laboratory for 'Culture and Sensitivity Testing'. Relationships with the IDI laboratory pre-dated the MSI. We were aware that IDI had found it very difficult to persuade local health workers to take swabs and samples for testing. A presentation by the IDI team to the IPC Committee in 2018 noted the absence of microbiology testing until 2016. The significant refurbishment of the laboratory in 2018, coupled with capacity-building, improved power supplies and a constant supply of reagents and general supplies created new opportunities. Despite this situation, the presentation concluded that: *There is very limited utilisation of the microbiology laboratory at FPRRH.*

A subsequent presentation in 2019, and despite significant attempts by the laboratory team to raise awareness on the wards through short training sessions, echoed this concern:

There is low up take of available microbiology diagnostics at FPRRH.

Prior to the commencement of the intervention, most of the patients in FPRRH were prescribed antibiotics on the basis of experiential knowledge (of clinicians). This is influenced by prior education, reference to clinical guidelines and heavily circumscribed by what is available, and perceptions of what is available, in pharmacy stores. This process is termed 'empirical therapy'.[2] There is increasing evidence that many of the most commonly used (and cheaper) antibiotics available in low resource settings (including public hospitals in Uganda) are no longer effective in treating infections; the bacteria have become resistant to them. Continued and extended use of these antibiotics on resistant infections will fail to deliver positive patient outcomes and waste antibiotics, contributing to AMR and hospital costs. Where, as in many RRHs in Uganda, there is no capacity for laboratory testing, doctors have little choice but to throw what is available at

---

[2] Empirical therapy refers to, 'Treatment given without knowledge of the cause or nature of the disorder and based on experience rather than logic. Sometimes urgency dictates empirical treatment, as when a dangerous infection by an unknown organism is treated with a broad-spectrum antibiotic while the results of bacterial culture and other tests are awaited' https://medical-dictionary.thefreedictionary.com/empirical+treatment.

patients. The use of microbiology testing in the hospital is beneficial both for individual patient management (ensuring they get the optimal antibiotic for the infection they have) and for creating evidence-based policies to guide empirical antibiotic use.

BACTERIOLOGY-BASED (INDIVIDUALISED) PRESCRIBING

The priority for the MSI, in the first instance, was to create the opportunity for individual rational prescribing based on culture and sensitivity testing. From the point of taking a wound swab or blood sample, it can take 5–7 days for antibiotic susceptibility results to be returned to the ward. Taking this into account, alongside the urgency of treating sepsis without delay, many women are prescribed an initial empirical antibiotic regimen until test results are received and communicated to a prescriber. In the Ugandan context, this means waiting for a doctor to reassess the patient and prescription. And it could take many days for the same doctor to review that case.

FACILITATING CULTURE AND SENSITIVITY TESTING

The qualitative interviews with health workers on the PNG ward indicate the priority that they all now ascribe to culture and sensitivity testing. Indeed, it is the normalisation of this process, through substantial and sustained behaviour change, that has underpinned the emergence of highly effective team working. And this, in turn, lies at the heart of intervention success (discussed below). The following midwife describes how this change has happened and the complex chain of events involved:

> If we see pus, we do a swab – but those things have just come.
> [So, you were not swabbing before?]
> We were not taking swabs – because if you took a swab it stayed in the laboratory for a month and there was no one to follow up. Patients stayed with us for over a month and there was no follow up at all. Those people (K4C midwives) do follow up and within 3 days they have results. They bring them to the doctor, so the doctors are ready to work on them and that's how we have reduced sepsis on the wards. It's great indeed I'm very happy.
> [So, before the project – was it that there was trouble in the laboratory or that no one here was picking the results from the laboratory – was it a communication problem?]

Yes at first we sent the swabs there and it could stay there 2 weeks with no one to follow up the swabs – the laboratory had a problem - they could tell you we don't have things to use –'you wait' – but nowadays things are there to use – you go and they bring back the results when they are out.

[Does having the phone on the ward help?]

Yes, now we use the phone that was bought for us to call the laboratory and also to receive results – it is TOO GOOD! Before we had the phone, we had to move from our ward to the laboratory - if we wanted to speak to the laboratory or take a sample we had to walk up there.

[And did people do that?]

No - some did, and others refused especially when we are alone on duty. So to do all those things – you are tired - you are alone – now [K4C midwife] has come, she can go to the laboratory and they are helping with dressing so for you – you are giving treatment – you are recording in the files – the work now is easy.

[Is it only K4C midwives contacting the laboratory? If they were not there would you do that?]

Yes, these days I'm used to it. We go there – we get the results – we call the doctor – we have the phone - Now we can call a doctor and get immediate follow-up.

[What happens when the results are back?]

We give the results to the doctor – he is the one to decide if they have resistance or not and if they have resistance to some antibiotics they change – if not they continue.

Although she refers to earlier experiences and perceptions of laboratory functionality, there are also issues concerning leadership and roles in terms of taking swabs in the first place and ensuring these physically get to the laboratory without delay. To put this in context, the laboratory is about a 10-minute walk from the ward and often there are only two midwives working on the ward. Leaving critically ill patients for a period of about 30 minutes is a real problem. This is another example of the mundane, minutiae of operational practice on the ward captured through continual observation and dialogue that would be missed using many other research methods and yet indicated a significant systemic weakness and facet of behaviour. The process of taking the swabs to the laboratory has now been streamlined with the laboratory collecting samples after the morning ward round. This also provides an opportunity to engage with the team on the ground. Where necessary, the midwives will physically take swabs to the laboratory during the day. This has been facilitated by the additional human resource provided through the project.

We also found at an early stage in the project that there was a major problem with the communication of results between the laboratory and the ward. The laboratory was attempting to contact the doctor who originally signed the swabbing forms. In practice, the presence of senior doctors on the ward is, at best, sporadic and it is intern doctors who are present for signing. But intern doctors inevitably move around and rotate. Guided by previous experience of an initiative to co-design and implement maternal early warning scores in Mulago Hospital (Ackers and Ackers-Johnson 2016), we were aware that reliance on the personal mobile phones of these doctors and health workers, who receive no funding to cover phone costs, leads to severe communication breakdown. We were concerned too at the impact of having midwives leave the ward perhaps several times a day to physically visit the laboratory for results. On that basis, the project supplied a landline in the nursing station to facilitate smooth and efficient laboratory-ward communication. This initially worked well but on a later visit we found the phone in a cupboard as some staff had been using it for other calls. We later provided fixed housing for the phone on the desk and a book to record calls.[3]

A representative of the laboratory echoes the midwife's perception of improved behaviour change in swabbing and prescribing practices.

> I have seen a very great improvement. Yes, there are bottlenecks, but we have really achieved a lot. Restricting our discussion to post-natal; the utilisation of diagnostics in regard to microbiology previously was very low. The staff didn't appreciate the value of prescribing on the basis of a laboratory report. Prescribing practices were poor; how often did the doctors come to check patients? They were wasting time. People believed we can use antibiotics the way the Ministry has put them into the clinical guidelines but some of these are not working so they really need to be guided by the bacterial reports. The uptake is really good. The number of samples has increased. Patients are being investigated along the lines of bacteriology – they really endeavour to pick results and the results are directly influencing prescribing behaviour. Although it's not been documented, the reactions from the nurses give us good news. The patient stay on the ward is actually reducing. We evidenced this last year when we monitored some few patients who had stayed up to 30 days on the ward without getting a bacterial report. From the time they prescribed on the basis of the bacteriology

---

[3] The lack of land lines in Ugandan health facilities is a major factor inhibiting communication both with patients and other health facilities.

report it took only 7 days for the patient to improve and finish treatment. On the 8th day they were able to close the wound and on the 9th the patient was discharged. This was during the project – last September. We have worked with the K4C team to do CMEs on IPC and swabbing. The main impact has been on rational prescribing – on a scale of 1-10 I'd say we are now at 7.8! It has helped us to inform the clinicians on best practice in terms of prescription and having done that we have reduced patient stay on the ward and then even in terms of costs – shorter stays on the wards and having fewer attendants on the wards.

The impact of the focus on wound care and culture and sensitivity testing is explained by a local midwife. She had taken a particular interest in the use of sugar in wound care prior to the project (in Sudan) and had previously worked alongside K4C staff and British nursing students on the labour ward, so relationships were strong. She describes the impact the project has had on her personally and on the ward and patients. She notes that, prior to the project, empirical prescribing of antibiotics was not effective, and this lack of effectiveness was compounded by prolonged prescribing of the same antibiotics. Importantly, she also specifically recognised the role that clinical pharmacists are now playing:

You came in at a critical time [and] brought new skills. Before there was no culture and sensitivity testing. Some of us knew about it but had never used it – even the doctors. When you came in it is me who benefitted most; I was carrying a very heavy burden and you helped me. You came as a combined team. We have not lost any women from sepsis since the project started. [Ugandan midwife employed by K4C] came. I had worked with her on labour ward with your students. Even the laboratory has started to respond – the burden was lifted, and everyone started getting involved.

We did use culture and sensitivity tests in Mulago (National Referral Hospital) but with not much emphasis and sometimes you have your interests on other things and we left it to the doctors. Here much of the things are now done by nurses/midwives – like doing culture and sensitivity tests. We knew culture and sensitivity would get results. Now I try to do the septic patients first. Before we noticed some were not getting better and we did not pay much attention to how this woman has been on this treatment for so long and you just gave her more antibiotics. [...] now [the pharmacist] comes on the ward daily and looks around and helps us as sometimes the intern doctors are busy and lack supervision. Before we used the same medicines – same – same – we just gave what was prescribed.

The Global Point Prevalence Survey completed in May 2019 monitored the use of microbiology testing in prescribing behaviour. At this point, prior to project engagement, no laboratory results were recorded in the files of those patients in the gynaecology or post-natal wards who had been prescribed antibiotics and 100% of antibiotic prescribing was on an empirical basis (Table 5.1).

We have previously noted the remarkable and very rapid improvement in wound management and swabbing once the MSI began to engage. The FPRRH laboratory has excellent documentation systems in place. The results from the laboratory indicate major behaviour change on the ward with 95% of all suspected sepsis cases tested once the project was implemented (Table 5.2).

Table 5.1 Antibiotic prescribing based on laboratory results

	Gynaecology (n = 22)	*Post-natal* (n = 20)
No. of patients on antibiotics	10	18
Microbiology lab report in notes	0	0

Source Results of G-PPS, May 2019 as reported to FPRRH IPC Committee

Table 5.2 Volume and proportion of suspected sepsis cases sent for laboratory testing

Time frame	*Suspected sepsis cases*	*Culture and sensitivity tests performed*	*% Tested*
January 1st, 2019–July 8th, 2019	50	0	0
July 9th, 2019–July 21st, 2019	16	3	19
July 22nd, 2019–January 31st, 2020	76	74 (2 had missing data)	95

Source FPRRH Laboratory

Laboratory results from these tests were subsequently located in the files of 67 of the 74 (90.5%) patients who had had a swab taken.[4] Although this emphasises the need to further improve record-keeping, this level of documentation represents a remarkable achievement in the context.

Utilising Laboratory Data and Antibiograms to Improve Empirical Therapy

In addition to providing the opportunity for an individualised treatment regime, the submission of samples to the laboratory builds up a wider evidence base of resistance patterns specific to that hospital setting. This evidence base enables microbiologists to identify more general resistance patterns that determine the expected efficacy of individual antibiotics. The collection of a sufficiently robust evidence base enables the hospital to generate what is known as an 'antibiogram'. An antibiogram is a collection of data, based on laboratory testing of the pathogens in a specific facility that summarises patterns of resistance to different antimicrobial agents (or antibiotics). Although international and national trends in resistance patterns can be identified, regional and facility-specific patterns enable even closer targeting of antibiotics. The intern doctor who had moved from Mbale hospital (above) reported his perception of regional differences in resistance patterns:

> The challenge here is that bacteria are more resistant than in Mbale. Here bacteria are often resistant to more than one antibiotic. We have been culturing E.coli here and there is much resistance.

Where a hospital antibiogram exists, initial prophylactic and empirical prescribing can be informed by local evidence and has a much higher chance of success. The presence of an antibiogram with associated awareness raising and sensitisation amongst all staff and especially medical interns would have major impacts on empirical prescribing across the hospital. Not only can it help reduce the reliance on and overuse of

---

[4] For four patients, the results had gone missing from the file; for two patients, the test was not completed because the IDI laboratory was closed over Christmas and New Year and one patient's lab test was not completed because the patient had discharged herself against medical advice.

specific antibiotics, a driver of resistance, it can act as a guide for procurement of more potent antibiotics at the hospital. The MSI has played an important part in creating the evidence base for a hospital antibiogram. Prior to the MSI, FPRRH did not have the volume of laboratory results to create the necessary evidence base for a hospital antibiogram. A member of the pharmacy team describes how this has changed:

> If you go to maternity, you will notice a very big change. The ward sends the biggest volume of swabs now to the laboratory because those people [midwives] are aware.

The laboratory scientist confirms this:

> On the basis of the increased swabbing we hope to be in a position to have an antibiogram. This will be very informative. The sample size is now very adequate. The antibiogram will be good for the clinicians to guide prescribing and it will be good for the patients.

He goes on to refer to the contribution that swabbing has made to an understanding of health care acquired infection and IPC monitoring. That week had seen 3 cases of *Acinetobacter*, a multidrug-resistant organism on PNG ward:

> Having the bacteriology reports coming through has also helped us. We want to share a report that will be coming next week on the trends of pathogens being identified and sensitivity patterns to see if this is the same organism from one patient to another so we can have targeted interventions when it comes to IPC. We can also see if it is airborne or a contagious organism and then we can improve on the IPC. Now we don't have that good background on IPC. It has helped us to do localised epidemiology on hospital acquired infections.

Not only has the practice of sending samples to the laboratory improved dramatically over the period of the project, the laboratory reports that PNG is responsible for nearly all of the laboratory testing that is taking place in the hospital and contributing to the forthcoming antibiogram. The specific impact of team working with IDI has underpinned all of these developments:

> [Have you noticed more swabs being taken from other wards?]

Table 5.3 Summarised data for microbiology testing on samples drawn from post-natal and gynae wards, FPRRH (July 2019–January 2020)

Sample type	No. Samples	Growth	No growth	Contamination
Pus swabs/Aspirates	82	52	28	2
High vaginal swabs	16	10	6	0
Blood	16	1	15	0
CSF	1	0	1	0
Urine	1	0	1	0
Total	116	63	51	2

Source FPRRH/IDI Laboratory

Actually no – it is mainly post-natal and surgical (IDI has had a project on surgical). IDI cover the whole hospital, but we have had some real challenges persuading people of the value of prescribing on the basis of laboratory results. Some people don't have the time to really investigate these patients – some base prescribing on clinical guidelines. OK, but they see what is in those guidelines is no longer working. We are coming up with a highly resistant Acinetobacter and we should not be prescribing certain antibiotics – all these forces will come if we have the right information.

In the time since this interview took place, an operational antibiogram has been produced by laboratory staff at FPRRH, shown in Table 5.11.

This data set, drawn from the laboratory, includes wound swabs and other samples illustrating the more extensive growth in sampling on the ward. Due to laboratory capacity constraints commonly seen in lower-income countries, samples are only grown and tested aerobically, meaning the possible presence of any anaerobic bacteria[5] is missed. Of the 116 total samples sent to the laboratory, 63 (54%) reported positive growth[6] and 51 (44%) reported no growth. Only 2 samples had probable contamination highlighting the success of laboratory protocols (Table 5.3).

A single sample may be attributed to a single bacterial isolate, or less commonly, multiple bacterial isolates indicating a polymicrobial infection. This explains the higher number of isolates reported in

[5] Bacteria that thrive in the absence or minimal presence of oxygen.

[6] Indicating that bacteria are present.

Table 5.4 Number of isolates of each bacterial species identified from the samples tested

Isolate ID	No. of isolates
Escherichia coli	26
Acinetobacter species	17
Staphylococcus aureus	15
Klebsiella species	13
Enterococcus species	7
Coagulase Negative Staph	4
Proteus mirabilis	4
Streptococcus pyogenes	2
Streptococcus agalactia	1
Pseudomonas aeruginosa	1
Providencia stuartii	1
Candida species	1
Raoultella ornithinolyticus	1
Total isolates	93

Table 5.5 Antibiotic resistance patterns of *Acinetobacter species* isolated from PNG ward

Acinetobacter species (n = 17)

Antibiotic agent	Susceptible	Intermediate	Resistant	Resistant (%)
Imipenem	1	1	15	88
Cefepime	0	1	16	94
Cefotaxime	0	1	15	88
Trimethoprim/sulfame	2	5	10	59
Doxycycline	11	1	5	29
Ciprofloxacin	2	0	15	88
Amikacin (*n* = 9)	8	0	1	11
Piperacillin/Tazobactam (*n* = 9)	0	1	8	89
Gentamicin	1	0	16	94

Source FPRRH Laboratory

Tables 5.4, 5.5, 5.6, and 5.7 relative to Table 5.3. Culture identification (Table 5.4) and antibiotic sensitivity testing were performed for all positive isolates[7] grown. The most common organism isolated was *Escherichia*

---

[7] The term isolate refers to the isolation of a particular bacterial colony from amongst others on the agar plate.

Table 5.6 Antibiotic resistance patterns of *Escherichia coli* isolated from PNG ward

Escherichia coli. (n = 26)

Antibiotic agent	Susceptible	Intermediate	Resistant	Resistant (%)
Imipenem (*n* = 25)	25	0	0	0
Gentamicin	19	1	6	23
Chloramphenicol	22	1	3	11
Ampicillin	2	0	24	92
Cefotaxime	5	1	20	77
Ciprofloxacin	17	1	8	31
Doxycycline (*n* = 23)	2	3	18	78
Piperacillin/Tazobactam (*n* = 19)	16	1	2	11
Trimethoprim/sulfame (*n* = 25)	2	0	23	92
Cefuroxime	4	1	21	81

Source FPRRH Laboratory

Table 5.7 Antibiotic resistance patterns of *Staphylococcus aureus* isolated from PNG ward

Staphylococcus aureus (n = 15)

Antibiotic agent	Susceptible	Intermediate	Resistant	Resistant (%)
Chloramphenicol	14	0	1	7
Gentamicin	14	0	1	7
Cefoxitin (*n* = 13)	11	0	2	15
Trimethoprim/sulfame	9	1	5	33
Ciprofloxacin (*n* = 14)	11	0	3	21
Clindamycin	15	0	0	0
Erythromycin (*n* = 13)	6	1	6	46

Source FPRRH Laboratory

coli (28%), followed by *Acinetobacter species* (18%) and *Staphylococcus aureus* (16%). This corresponds with a study of 314 surgical site infections at Mulago National Referral Hospital, where 23.7% of isolates were *E. coli*, 21.1% were *S. aureus* and 17.1% were *Acinetobacter species* (Seni et al. 2013). The resistance patterns of these bacteria are further explored in Tables 5.4, 5.5, 5.6, and 5.7.

The antibiotic susceptibility of isolates was tested utilising the disk diffusion technique against a specific panel of antibiotics to best determine its resistance potential. The two antibiotics most commonly prescribed in the PNG ward: ceftriaxone and metronidazole, are not specifically mentioned in the tables below. These antibiotics are often prescribed together prophylactically prior to any laboratory testing. Metronidazole is commonly used to treat anaerobic bacteria (Smith 2018; Shafquat et al. 2019) and as such testing for this antibiotic requires laboratories capable of simulating anaerobic conditions. This is not possible at FPRRH. Ceftriaxone, on the other hand, is a member of the cephalosporin group of antibiotics. In this case, other members of the same family (such as cefepime or cefotaxime) with similar mechanisms of action can be used to infer resistance.

Table 5.5 evidences an alarmingly high level of resistance across all antibiotics tested against the 17 *Acinetobacter* isolates with the exception of doxycycline and amikacin (5 and 1 resistant isolates, respectively). That there were no isolates susceptible to cefepime or cefotaxime, fourth- and third-generation cephalosporins, respectively, is cause for concern. Not only are these some of the most recent iterations of antibiotics, but the fact that the primary antibiotic of choice on the ward is ceftriaxone shows *Acinetobacter* infections could leave people vulnerable. Equally, once an infection is present, few other antibiotics are shown to be effective.

The most prevalent bacterial species identified from the tests was *E. coli*. Table 5.6 shows that *E. coli* displays mixed levels of resistance, leaning towards being either highly susceptible or highly resistant depending on the antibiotic of choice. Again, there is high resistance to cefotaxime (20/26) as well as cefuroxime (21/26), a second-generation cephalosporin, which adds to the concern that ceftriaxone is losing its effectiveness. Thankfully, imipenem[8] has been shown to be 100% effective against the isolates tested, promoting its use as a secondary option.

The third most prevalent isolate was *S. aureus.* (Table 5.7) which showed minimal resistance to all antibiotics with the exception of erythromycin and trimethoprim/sulfame. Additionally, clindamycin was shown to be 100% effective against the isolates tested. Of particular note is cefoxitin which performed strongly with 11 susceptible and 2 resistant isolates. Cefoxitin is a second-generation cephalosporin and acts as

---

[8] We note the increased ordering of meropenem in the hospital procurement plan in Table 6.1.

an indicator for MRSA (methicillin-resistant *S. aureus*), a key metric for AMR.

Utilising Local *Staphylococcus* *Aureus* Populations as a Case Study to Investigate Trends in Resistance Patterns

Carried by roughly 30% of the human population, *Staphylococcus aureus* is a Gram-positive coccus found frequently in the nasopharynx, respiratory tract and on skin. Whilst commonly found as a commensal[9] organism, *S. aureus* is also a major human pathogen with the ability to cause a wide range of clinical infections resulting in skin (where the skin has been broken, for example, from a wound or surgery) and respiratory diseases (Tong et al. 2015). Infections occur most regularly in hospitalised patients where the consequences can be severe. Though often easily treated with antibiotics, if left neglected the infection can worsen and spread. *S. aureus* is one of several organisms commonly linked to bloodstream infections and cases of sepsis. A large meta-analysis of community acquired bloodstream infections in Africa examined 5578 patients with non-malaria bloodstream infections, where 531 (9.5%) cases were due to *S. aureus* (Reddy et al. 2010). That being said, treatment has become increasingly complicated due to the rise of methicillin-resistant *S. aureus* (MRSA) which is now often multi-resistant. As such, methods of infection prevention are becoming more valuable (Kluytmans et al. 1997).

With a better understanding of where and how resistance arises, steps can be taken to prevent it. Ongoing research forming part of Ackers-Johnson's microbiology doctorate (Ackers-Johnson 2020) investigates in depth the mechanisms of antibiotic resistance and their respective relation to different strains of *S. aureus*. As part of this, *S. aureus* isolates obtained clinically from across all wards at FPRRH were investigated (Table 5.8), as well as isolates obtained from the hands of healthy members of the public (Table 5.9)—all within the hospital catchment area.

As previously noted, production of an accurate antibiogram requires a large number of samples. Whilst the antibiotics' effectiveness relative to each other remains largely similar, a greater percentage of resistant isolates can be seen when assessing the 70 samples acquired from across

---

[9] Can live naturally on human skin without causing harm.

Table 5.8 Antibiotic resistance patterns of *Staphylococcus aureus* from all wards at FPRRH

Staphylococcus aureus (n = 70)

Antibiotic agent	Susceptible	Intermediate	Resistant	Resistant (%)
Chloramphenicol (*n* = 59)	53	0	6	10
Gentamicin (*n* = 68)	54	2	12	18
Cefoxitin (*n* = 21)	13	0	8	38
Trimethoprim/sulfame (*n* = 57)	25	6	26	46
Ciprofloxacin (*n* = 69)	51	4	14	20
Clindamycin (*n* = 62)	62	0	0	0
Erythromycin (*n* = 69)	47	1	21	30
Tetracycline (*n* = 53)	24	0	29	55

Source These bacteria were clinically isolated from samples across all wards at FPRRH between August 2017 and April 2019. Whilst 70 isolates were assessed, not all could be tested against all antibiotics due to supply constraints as the laboratory was in its initial stages of increasing capacity

Table 5.9 Antibiotic resistance patterns of *Staphylococcus aureus* isolated from hand swabs of the general public

Staphylococcus aureus (n = 125)

Antibiotic agent	Susceptible	Intermediate	Resistant	Resistant (%)
Chloramphenicol (*n* = 124)	87	0	37	30
Gentamicin (*n* = 123)	116	0	7	6
Cefoxitin	98	0	27	22
Trimethoprim/sulfame	54	4	67	54
Ciprofloxacin	81	0	44	35
Clindamycin (*n* = 124)	24	53	47	38
Erythromycin (*n* = 124)	39	28	57	46
Tetracycline	57	31	37	30

Source Ackers-Johnson (2020)

the hospital (Table 5.8) compared to the 15 samples from the PNG ward (Table 5.7). In the case of gentamicin and cefoxitin, the rates of resistance are more than doubled. Interestingly, even when observing the larger sample size, clindamycin has still been shown to be 100% effective.

Observing the resistance profiles of *S. aureus* found on the hands of healthy members of the public within the catchment area of FPRRH highlights how AMR is not a problem reserved exclusively to hospitalised patients. Table 5.9 shows relatively high levels of resistance across all antibiotics tested with the exception of gentamicin at 6%. Moreover, clindamycin, which was shown to be 100% effective against isolates obtained clinically at FPRRH, has much higher levels of resistance within the local community at 38%. With such levels of resistance present, the requirement for effective IPC is further emphasised to reduce the incidence of infection from harmful resistant bacteria already present on skin. This raises questions about the role that attendants may play as vectors of infection on the wards especially when they are involved in clinical roles (wound and canula management) and the impact of stock-outs of iodine on surgical safety practices in theatre.[10] Hsieh et al. (2014) generated a theoretical model to predict the impact of hospital visitors on nosocomial transmission and spread to the community concluding that, 'transmission rates of infective residents in the community and of infective visitors at the healthcare facility have a decisive impact on disease eradication/persistence' (2014: 20). These concerns are emerging in current research on COVID-19 (Halbfinger 2020; Hsu et al. 2020; Wee et al. 2020).

Comparing the situation at FPRRH to that of the National Referral Hospital (Mulago Hospital, Table 5.10), the results again seem relatively similar with trimethoprim/sulfame exhibiting exceedingly high levels of resistance, with gentamicin and chloramphenicol being amongst the strongest performing antibiotics. Conversely, of particular note is clindamycin. 40.6% of *S. aureus* isolated from surgical site infections at Mulago hospital displayed resistance to clindamycin, similar to that of the community hand swabs in Fort Portal (38%), whilst clinical isolates from FPRRH were 100% susceptible. Such data emphasises the need for further advanced research.

Tables 5.5, 5.6, 5.7, 5.8, and 5.9 highlight the necessity of microbiology testing in controlling infection and promoting antimicrobial stewardship. It is clear from the data that there is no antibiotic 'silver bullet' available in this local setting capable of treating an infection of unknown identity. All the bacteria assessed at FPRRH had some level of

---

[10] At one point during stock-outs, staff had only saline to use.

Table 5.10 Antibiotic resistance patterns of *Staphylococcus aureus* clinically isolated from surgical site infections at Mulago Hospital

Staphylococcus aureus (n = 64)	
Antibiotic agent	*Resistant (%)*
Chloramphenicol	15.6
Gentamicin	18.8
Ampicillin	100
Trimethoprim/sulfame	89.1
Ciprofloxacin	29.7
Clindamycin	40.6
Erythromycin	46.9
Tetracycline	42.2

Source Seni et al. (2013)

resistance to all antibiotics tested, except for imipenem (100% effective against *E. coli*) and clindamycin (100% effective against *S. aureus*). As such, prescribing without the respective laboratory data could well prove to be in vain. The unchecked overuse of these ineffective antibiotics not only compromises patient safety and acts as an economical burden, but also further drives the development of resistance—with increased levels of antibiotics in the local environment it is possible that other bacteria unrelated directly to the cause of infection can progress to acquire resistance causing issues in the longer term. The construction of the antibiogram (Table 5.11) will go a long way to address these problems, the reliability of which itself has been greatly aided by the increased volume of samples being taken as a by-product of effective patient care.

The antibiogram displays the percentage susceptibility of isolates recovered from microbiological samples analysed at FPRRH. The first column indicates the organisms that were assessed, with the remaining columns indicating the antibiotics that they were tested against, the percentage of isolates that were susceptible to the antibiotic, and (n)—the total number of isolates tested. It is worth noting that data where fewer than 30 isolates have been assessed are less reliable and hence any conclusions/prescribing should be done with caution. These have been marked with an asterisk (*) and highlighted yellow.

As previously mentioned, a wide range of resistances to various antibiotics can be seen across the data. As a general rule, if an antibiotic has been shown to be less than 80% effective against a particular organism,

Table 5.11 Hospital Antibiogram (July 2020) showing percentage susceptibility of isolates based on data collected from January 2019 to June 2020 at Fort Portal Regional Referral Hospital

	Aminoglycosides		Beta-lactam combination		Beta-Lactams / Cephalospor Ins			Carbapenem	Penicillins		Fluoroquinolones	Macrolides		Folates	Phenicols	Glycopeptides	Lincosamides	Tetracyclines	Cephamycin
	GEN % (n)	AMK % (n)	TZP % (n)	AMC % (n)	CTX % (n)	CEP % (n)	CRX % (n)	IPM % (n)	AMP % (n)	PEN % (n)	CIP % (n)	ERY % (n)	AZM % (n)	SXT % (n)	CHL % (n)	VA % (n)	DA % (n)	DOX % (n)	FOX % (n)
Gram Negatives																			
Acinetobacter spp	27.2 (81)	88.5 (26*)	29.7 (37)		9.6 (83)	28.4 (74)		43.0 (86)			25.8 (85)			27.1 (85)				77.6 (67)	
E. coli	62.2 (196)	94.0 (67)	79.1 (91)	31.7 (167)	32.4 (201)	22.3 (184)	19.7 (198)	99.0 (199)	4.0 (174)		38.3 (201)			13.6 (199)	71.3 (181)			16.9 (148)	
Citrobacter spp	72.2 (18*)	100.0 (5*)			55.0 (20*)	58.8 (17*)		100.0 (20*)			55.0 (20*)			50.0 (20*)	70.0 (20*)			69.2 (13*)	
Pseudomonas aeruginosa	100.0 (21)	100.0 (10)	83.3 (12)			100.0 (17)		100.0 (20*)			100.0 (21)								
Klebsiella spp	54.1 (98)	97.3 (37)	88.0 (50)	46.6 (88)	31.1 (103)	33.0 (91)	32.0 (103)	96.2 (104)			62.5 (104)			19.2 (104)	57.0 (100)			40.6 (69)	
Proteus vulgaris	81.3 (16*)	100.0 (2*)	100.0 (11*)	81.3 (16*)	68.8 (16*)	73.3 (15*)		81.3 (16*)			68.8 (16*)			25.0 (4*)	75.0 (16*)			43.8 (16*)	
Proteus mirabilis	42.1 (38)	100.0 (3*)	100.0 (26*)	70.0 (10*)	22.9 (35)	20.8 (26*)	18.4 (38)	97.4 (38)	16.1 (31)		26.3 (38)			10.5 (38)	61.1 (36)			3.0 (33)	
Gram Positives																			
Enterococcus spp									75.0 (20*)	75.0 (20*)	22.7 (22*)	18.2 (22*)				85.0 (20*)			
Staph aureus	80.5 (113)										74.1 (116)	64.3 (115)	66.1 (115)	69.0 (113)	90.3 (113)		99.1 (114)	70.9 (86)	63.8 (116)

GEN-Gentamycin, AMK-Amikacin, TZP-Piperacillin/Tazobactam, AMC-Amoxicillin/Clavulanate, CTX-Cefotaxime, CEP-Cefepime, CRX-Cefuroxime, IPM-Imipenem, AMP-Ampicillin, PEN-Penicillin, CIP-Ciprofloxacin, ERY-Erythromycin, AZM-Azithromycin, SXT-Sulphamexazole/Trimethoprim, CHL-Chloramphenicol, VA-Vancomycin, DA-Clindamycin, DOX-Doxycycline, FOX-Cefoxitin (n)-Number of isolates, %-Percentage of susceptibility,
*Results based on fewer than 30 isolates are less reliable and should be interpreted with caution

then it should not be prescribed as an empirical therapy for serious infections. As per the antibiogram, a number of antibiotics fall short of this benchmark. With relatively more expensive and less commonly prescribed antibiotics (e.g. Amikacin and Imipenem) often outperforming those that are frequently used, it is important that the antibiogram plays a role in the hospitals' antibiotic procurement plan as particular treatments are simply not health or cost-effective. Whilst budget constraints may not be able to permit the frequent use of such tailored antibiotics, clear policies and guidelines should be put in place by the MTC to govern their usage in the care of critical patients.

Whilst there is a definite value in utilising an antibiogram, it is also useful to point out that it has limitations. Being used for empirical diagnoses, there are no guarantees that the antibiotic of choice will be effective even if it has a very high theoretical rate of success. This could be due to the performance of the antibiotic (and antibiotic quality) or the (mis)identification of the causal bacteria. As such, the presence of an antibiogram does not remove the essential need for clinicians to utilise the laboratory services available to provide definitive information in conjunction with patient observations to deliver optimal treatment.

The growth in wound management coupled with increased laboratory testing has created task-shifting opportunities (and workloads) for nurses and midwives as staff providing continuity of care on the wards. It has also created new opportunities for the engagement of clinical pharmacy. Chapter 6 discusses these issues in more depth and then assesses one of the major, structural, barriers to further improvement: supply chain dynamics in Ugandan public hospitals.

References

Ackers, H. L., & Ackers-Johnson, J. (2016). *Mobile professional voluntarism and international development: Killing me softly?* Palgrave PIVOT. http://link.spr inger.com/book/10.1057%2F978-1-137-55833-6.

Ackers, H. L., Ackers-Johnson, G., Seekles, M., & Opio, S. (2020). Opportunities and challenges for improving anti-microbial stewardship in low- and middle-income countries; lessons learnt from the maternal sepsis intervention in Western Uganda. *Antibiotics, 9,* 315. https://doi.org/10.3390/antibioti cs9060315.

Ackers-Johnson, G. (2020) *Comparing the antimicrobial diversity of Staphylococcus aureus strains isolated from clinical cases of infection and those found as a commensal organism in Fort Portal, Uganda and further investigating the potential mechanisms of resistance present* (PhD research on-going).

Halbfinger, D. M. (2020, April 21). Hospitals in Israel let relatives say goodbye to loved ones up close. *New York Times*.

Hsieh, Y.-H., Lui, J., Tzeng, Y.-H., & Wu, J. (2014). Impact of visitors and hospital staff on nosocomial transmission and spread to community. *Journal of Theoretical Biology, 356,* 20–29.

Hsu, Y.-C., Liu, Y.-A., Lin, M.-H., Lee, H.-W., Chen, T.-J., Chou, L.-F., et al. (2020). Visiting policies of hospice wards during the COVID-19 pandemic: An environmental scan in Taiwan. *International Journal of Environmental Research and Public Health, 17,* 2857.

Kluytmans, J., van Belkum, A., & Verbrugh, H. (1997). Nasal carriage of Staphylococcus aureus: epidemiology, underlying mechanisms, and associated risks. *Clinical Microbiology Reviews, 10*(3), 505–520.

Reddy, E. A., Shaw, A. V., & Crump, J. A. (2010). Community-acquired bloodstream infections in Africa: A systematic review and meta-analysis. *Lancet Infect Dis., 10*(6), 417–432. https://doi.org/10.1016/S1473-3099(10)700 72-4.

Seni, J., Najjuka, C. F., Kateete, D. P., Makobore, P., Joloba, M.L., Kajumbula, H., et al. (2013). Antimicrobial resistance in hospitalised surgical patients: A silently emerging public health concern in Uganda. *BMC Research Notes, 6,* 298.

Shafquat, Y., Jabeen, K., Farooqi, J., Mehmood, K., Irfan, S., Hasan, R., et al. (2019). Antimicrobial susceptibility against metronidazole and carbapenem in clinical anaerobic isolates from Pakistan. *Antimicrob Resist Infect Control, 8,* 99. Published 2019 June 14. https://doi.org/10.1186/s13756-019-0549-8.

Smith, A. (2018). Metronidazole resistance: A hidden epidemic? *British Dental Journal, 224*(6), 403–404. https://doi.org/10.1038/sj.bdj.2018.221.

Tong, S. Y., Davis, J. S., Eichenberger, E., Holland, T. L., & Fowler, V. G. (2015). Staphylococcus aureus infections: Epidemiology, pathophysiology, clinical manifestations, and management. *Clinical Microbiology Reviews, 28*(3), 603–661.

Wee, L. E., Conceicao, E. P., Sim, X., Y. J., Aung M. K., Tan, K. Y., Wong, H. M., et al. (2020). Minimizing intra-hospital transmission of COVID-19: The role of social distancing. *Journal of Hospital Infection, 105,* 113–115.

CHAPTER 6

Task Shifting, Midwifery Empowerment and the Nascence of Clinical Pharmacy

Abstract This chapter addresses the role that the intervention has played in shaping professional engagement within the multi-disciplinary team. The existence of laboratory results has triggered the emergence of clinical pharmacy roles. The chapter traces the impact of this on prescribing behaviour and on procurement planning and hospital policies. Whilst celebrating the progress made and viability of the model, it describes the structural impact that access to antibiotics and IPC supplies has on the realisation of optimal change.

Keywords Antimicrobial stewardship · Maternal sepsis · Task shifting · Clinical pharmacy · Pharmacotherapy · Antibiotic consumption · Antibiotic susceptibility · Procurement planning · Stock-outs

In addition to describing the importance that swabbing and testing have made to progress, the midwife in the previous chapter alludes to a major change in team working and task shifting with midwives and nurses now playing a very central role in these processes. This has been critical to the effectiveness of the MSI. Endorsed by the WHO (2007), task shifting has been implemented in the global health care setting in an effort to combat healthcare worker shortages, increase efficiency and cut costs. In general terms, task shifting is the delegation of a specific task to a lower cadre (WHO 2008). Utilising the available workforce

L. Ackers et al., *Anti-Microbial Resistance in Global Perspective*,
https://doi.org/10.1007/978-3-030-62662-4_6

in this manner promotes the more effective and efficient use of human resources. With greater numbers of healthcare staff being able to offer certain aspects of care or perform certain clinical procedures, it follows that there is an increased ability to provide healthcare services, coupled with improvements to healthcare worker skills, greater efficiency within the healthcare system and cost savings (WHO 2007). Task shifting in the West implies an increase in remuneration and substantial investment in continuous professional and career development. The reference to non-medical prescribing[1] is a case in point. The same cannot be said in LMICs where healthcare professionals are often burdened with additional responsibilities without being compensated accordingly. Nonetheless, task shifting is recommended as a method to reduce healthcare worker shortages in the maternity setting (WHO 2012). Despite a lack of policy framework to support task shifting in Uganda, it has been unofficially practised in the country since 1918 (Baine et al. 2018).

Whilst it is the responsibility of all healthcare professionals to act as antibiotic stewards, we have found that although midwives and nurses are barely mentioned in the NAP, they have, by means of informal task shifting, taken on a significant role in advocating for antimicrobial stewardship. In the Ugandan context, their active engagement and empowerment are absolutely essential to AMS, not least because they are most often the only cadres continually present on the ground. The presence of senior doctors is at best sporadic with rotating and largely unsupervised intern doctors providing most medical input (Ackers et al. 2016; Tweheyo et al. 2019). This evidence adds weight to Brink et al.'s proposal for new nurse-led models of AMS in Africa (2016).

A midwife respondent explains how the new knowledge, combined with team working and support, has facilitated effective task shifting on the ward:

> Before [the project] staff did not have the idea about wound swabbing; they didn't understand culture and sensitivity testing. That's where a lot of improvement has come. They now understand the science and take an interest and ask, 'why is this mother not getting better?' They have learnt about resistance. Midwives are now taking the lead in managing patients; they used to wait for doctors to make decisions, but the doctors only appear once in a while, so they waited for them instead of acting. Now

---

[1] See note 31.

they act. They used to say, 'that one is for the doctor – what has killed this mother?'

The intern doctors on the ward have also observed marked changes in the behaviour of the midwives—and their own practices. One intern doctor commented that, when he arrived:

> We found a lot of sepsis on the ward but during that period there was a scarcity of things to use – antiseptics, saline; then the staff started taking off cultures. You have really helped us. Your team has done a lot to reduce sepsis here. They assess the wounds – take off the culture – bring the results as to which microorganisms are resistant, so we cleared all septic mothers.
> [Has swabbing practice changed?]
> Yes, it has changed – the nurses[2] do it perfectly now.
> [Would you say most wounds are now being swabbed and going to the laboratory?]
> In fact, all of them. Once we get a septic mother they will be swabbed. Immediately the first thing is to do a pus swab.

Another intern doctor acknowledges that behaviour change has been evident amongst all cadres:

> I really think we did have that knowledge before, but we did not appreciate its significance
> [When you say 'we' do you mean the doctors or midwives?]
> I think you should consider all of us. We did not do as much as we should have been doing. As long as we dressed the wounds once a day and I prescribed antibiotics the rest of the things would take their course. So, what K4C has brought to reality is that – actually when this is done, and samples are taken and followed up then things get better more quickly.
> [You said before people were aware of culture and sensitivity testing – they had the knowledge but didn't appreciate its significance?]
> Exactly. [K4C midwives] take the swabs and we get results now in three days. They also get you the pharmacists on the ward so we can target that specific organism.

---

[2] It is common in Uganda for doctors to use the term 'nurse' generically to include midwives.

The intern doctor in the quote above notes the role that the midwives play in anchoring pharmacy into the MDTs. One of the most important changes brought about as a direct result of increasing wound management and culture and sensitivity testing has been the way that this has facilitated clinical pharmacy. Since the project commenced, pharmacy has been directly involved in patient management and prescribing decisions as part of a proactive multi-disciplinary team. It is important to note that the pharmacists were fully aware of the benefits of this approach prior to the project and had taken steps to actively encourage it. There was no evidence of a knowledge gap as such. However, until this became a reality on the ward, their ability to utilise their expertise in clinical pharmacy was limited. Chapter 1 reviewed predefined metrics set out by the funding body. Pharmacists on the ground were highly sceptical of the value of antibiotic consumption metrics in this environment stating that:

> We cannot measure performance by zeroing on antibiotic [consumption] only.

They proposed instead what they felt was a more relevant indicator, namely the 'Review of Pharmacotherapy' by pharmacists. We are concerned here with the extent to which pharmacists are directly engaged in multi-disciplinary decision-making following the receipt of laboratory test results. Prior to the MSI, it was rare to have pharmacists present on the ward. A K4C midwife explains the importance of the culture and sensitivity testing process to the harnessing of clinical pharmacy engagement:

> Pharmacy will tell you there is no point in them coming unless there are cultures.

The data collated from patient notes showed that, once the intervention was in place, pharmacists reviewed the pharmacotherapy in 91.8% of cases where test results were taken and showed a bacterial growth.[3] The regular presence of pharmacy on the ward has in turn formed a critical part of the mentoring and knowledge transfer amongst the whole team (and explicitly including the K4C staff and UK volunteers). One of the pharmacists

---

[3] Patient Records (January 2019–January 2020).

describes the changes in practice and the role that laboratory results have played in this:

> Culture and sensitivity testing has been very important especially in post – c-section sepsis. We need to know what we are dealing with. Antibiotic selection is a big issue in AMR. Before they were making a blind selection. They were using only the antibiotics they were used to, the standard arrangement of routine antibiotics. Patients were staying for an extended period of time; the drugs were not working. They never figured it out that they were using the wrong antibiotics. Now it has become standard practice that you take a sample and send it to the laboratory and wait for the results to come. We have now moved to individualised antibiotic selection based on the laboratory results. We know what [the infection] is susceptible to.

He goes on to outline the need for more support through informal mentoring and formal training to enable all cadres to interpret the laboratory reports:

> Most of the problems now are how to interpret and apply the [lab.] reports. It is important we all agree on the value of that report and how to interpret it and do things in accordance with the findings.

The project responded to this explicit request for training by organising a small session for staff on the PNG ward. Later, the pharmacy team took the initiative to organise a larger training workshop for representatives of the entire hospital at a morning meeting with intern pharmacists taking the lead.

The MSI has achieved optimal pharmacy engagement (in a RRH context) on the PNG ward. The impact of laboratory testing has played an important role in empowering pharmacy. This is evident in the new policy, initiated by pharmacy with strong support from the laboratory, of only permitting use of high-end antibiotics when laboratory test results are available:

> Clinicians are not allowed to change antibiotics now without cultures.

Discussions have taken place in the hospital's IPC committee to extend this policy to all wards illustrating the wider impact of the MSI on the hospital. The laboratory respondent welcomes this achievement which reflects the growing recognition of pharmacy in the hospital:

> The policy of only prescribing high end antibiotics to patients who have had culture and sensitivity testing has really worked; these antibiotics are being guarded jealously now. In fact, (the pharmacy team) are very strict on that. I really feel this could work on other wards. It is only working on post-natal ward at present because they have laboratory reports.

The extent to which microbiology results have been understood, valued and applied by midwives on the ward is exemplified in the following excerpt:

> After Christmas (2019) we didn't have many septic patients but a week after four of them had the same isolates with the same results. We wondered how come four people worked on separately can come in with the same isolates? They came from different places. How can we get to know the source? We took the swabs off immediately. We did discuss with the laboratory people, but we were concerned so we looked at it ourselves. Is it the hygiene of the patients or the surroundings? Are there other possibilities? We know it is a HAI; Acinetobacter is not an organism you can find outside the hospital. The challenge is it is airborne – we spoke to the pharmacist and the laboratory. Having 4 cases with the same isolate is more of an outbreak. Maybe the laboratory could swab surfaces, or we could fumigate?

Ultimately, the model that has evolved on PNG ward has demonstrated the potential to significantly reduce infection, improve prescribing practice, reduce antibiotic consumption and overcome some of the effects of AMR on patient outcomes. To that extent, significant progress has been made to put systems in place to facilitate evidence-based prescribing. Ultimately, the impact of these systems hinges on access to antimicrobials.

Strategic Objective Three (NAP) highlights the issue of Optimal Access to essential antimicrobials and states very clearly that, *'the major modifiable driver of AMR is the use of antimicrobial agents'* (p. 14). Sub-objectives focus on distribution and supply chain mechanisms.

Procurement and Supply Chain Dynamics in Regional Referral Hospitals

The following section outlines critical problems in the supply chain system influencing the availability of antibiotics in Ugandan Regional referral hospitals. These include:

1. Budget Constraints: the hospital may only order against a budget prescribed by the Ministry of Finance and held centrally by National Medical Stores (NMS).
2. Limited Options: the hospital can only order antibiotics from a prescribed catalogue which excludes many of the antibiotics indicated as necessary from laboratory results and present on the Essential Medicines and Health Supplies List for Uganda (EMHSLU).
3. Failure to deliver on orders: major and unpredictable discrepancies exist in what is ordered and what is delivered ('Order Fill Rates').
4. As a result of the above, most IPC supplies and antibiotics run out half-way through the bi-monthly supply cycle (Stock-Outs).

In Uganda, the funds for procurement of drugs and supplies in the public sector are highly centralised and inadequate. NMS procures and distributes supplies to health facilities based on a centrally allocated Annual Supplies Budget. Each hospital is required to produce an Annual Procurement Plan. Once agreed, this Plan is fixed and cannot be varied over the year reducing the opportunity for flexibility and responsiveness to the hospital laboratory results and any changes indicated by a future antibiogram. This budget is held by NMS. With the exception of private wards, it is not possible for a RRH to source supplies from elsewhere. NMS deliveries take place bi-monthly.

During the annual procurement process, the hospital may only order those items authorised by the EMHSLU. However, not all essential drugs feature in the NMS catalogue. A hospital pharmacist describes the situation as follows:

> As much as we may desire a certain antibiotic, we can't plan for it if it is not present in the catalogue. A case in point is Amikacin and Moxifloxacin.

Out of the nine antibiotics tested against *Acinetobacter* samples in the laboratory, only two, doxycycline and amikacin, showed greater levels of susceptibility than resistance. Given the much higher success rate of amikacin, it is paramount that the antibiotic can be obtained for cases of severe *Acinetobacter* infections where other avenues have failed.

When the project started, the procurement plan for the financial year 2018/2019 was already in place. At that time, the plan did not take account of antibiotic susceptibility as reported by the laboratory, resulting in a lack of high-end antibiotics required to treat resistant bacterial strains. The evidence generated through the MSI, supported by the engagement of the K4C pharmacist, contributed to the procurement planning process for 2020/2021. Table 6.1 presents an extract from the hospital's procurement plan to indicate key antibiotics used on the post-natal and gynae ward. It evidences significant changes in antibiotic ordering and consumption arising directly from the intervention, where antibiotics in red denote project-related increases and those in blue denote decreases. For now,[4] we draw the reader's attention to column 5 (average past consumption). This shows the predominance of a small number of antibiotics including amoxicillin, metronidazole, ciprofloxacin, ceftriaxone and chloramphenicol.

ANTIBIOTIC DISTRIBUTION AT WARD LEVEL IN FPRRH

When supplies arrive at FPRRH from NMS, they are taken to the main stores. At this point, supply data is recorded electronically in the online ordering system (Rx). When we first visited pharmacy stores at the start of the project, we found the Rx system unused. The Rx initiative is a national programme designed to improve supply chain management and provide an online database linking the hospitals to NMS. When functional, it would provide a systematic method of stock utilisation and recording providing high-quality data to the hospital. It was clear that the pharmacy team had the expertise to use the software but there was no printer ink available. Two donated printers sat unused as they had no ink and obtaining cartridges for these was very complex. This illustrates the problems of managing donated devices in LMICs especially when they are not integrated in any way to local supply chains. The hospital had ink in stock but for very different printers and could not provide the printer ink to the stores. In this instance, the MSI provided a printer and printer ink

---

[4] We return to discuss the impact of the MSI on Procurement Planning in due course.

Table 6.1 Extract from the 2020/2021 procurement plan (with bi-monthly figures) focusing on key antibiotics used on PNG wards[a] at FPRRH

	Unit	Price	2019/2020 plan	Past Av. consumpt.	2020/2021 plan	Bi-monthly cost
Amoxicillin 250g capsules	1000	5,4100	101	134	90	4,869,000
Ampicillin/cloxacillin capsules	100	14,400	30	42	50	720,000
Metronidazole tablets	1000	17,100	50	70	70	1,197,000
Chloramphenicol 250mg tabs	1000	10,0200	0	0	5	501,000
Ciprofloxacin 50mg tablet	100	10,500	70	118	70	735,000
Ceftriaxone 1g vials	1	1700	6000	6000	5000	8,500,000
Ampicillin/cloxacillin 250mg inj	50	33,000	0	2	1	33,000
Meropenem 500mg inj	100	71,700	2	5	2	143,400
Gentamycin 80mg inj	1	16,000	0	92	150	2,400,000
Ciprofloxacin IV 200mg	100	32,800	40	0	0	0
Metronidazole 500mg infus	50	33,000	100	217	150	195,000
Cefotaxime inj	1	800	7000	5833	7000	5,600,000
Penicillin Benzathene 2.4MU amp	1	13,400	1	767	0	0
Chloramphenicol 1g inj.	10	11,400	5	4	5	57,000
	50	80,200	0	0	5	401,000

[a]In this table, the column 'unit' shows the number of doses per unit as sold. 'Price' is the price per unit in Ugandan Shillings (1 USD = 3780 UGX, May 2020). The column '2019/2020 plan' shows the number of units that were ordered for delivery every other month in 2019, with 'past av consumption' detailing the average bi-monthly consumption of the past year for each antibiotic. The column '2020/2021' details the set number of units ordered every other month for this year, with the corresponding bi-monthly cost reported in the final column
Source Hospital procurement plan, 2020–2021

for the duration of the project to enable the Rx system to become part of the organisational culture in pharmacy stores. Since then, the hospital has employed a dedicated stock manager to work with this system freeing up the senior pharmacist to engage in clinical pharmacy roles. We have discussed what may seem a mundane supplies issue in some depth here as this is precisely the kind of problem that lies behind major systems failures in Uganda, the diseconomies of 'donations' and the ease with which organisations become dependent on donor support. It also illustrates one of the underlying causes of data quality problems.

The process of distributing supplies *within* the hospital is, unfortunately, not presently covered by the Rx electronic system. Instead, the in-patient pharmacy (located a short distance from the stores) orders from the stores. Individual wards then visit the in-patient pharmacy to requisition supplies, and this is recorded manually on forms and in a records ledger book (the HMIS Dispensing Log).

Figure 6.1 presents data obtained from the in-patient pharmacy on the distribution of oral antibiotics between the main hospital wards. We recorded this data for the months of January and February 2020 to gain a picture of the distribution of injectable and oral antibiotics over the 2-month supply cycle and put PNG ward consumption patterns in a whole-hospital context. As can be seen, the level of antibiotic consumption on the PNG wards as a proportion of overall consumption is high and indicates the importance of this to overall antimicrobial stewardship:

Tables 6.2 and 6.3 identify antibiotic consumption on the PNG wards in more detail over a 4-month window. Table 6.2 provides data for Intravenous Antibiotics and evidences the predominance of metronidazole and ceftriaxone. Table 6.3 provides data for Oral Antibiotics and shows the

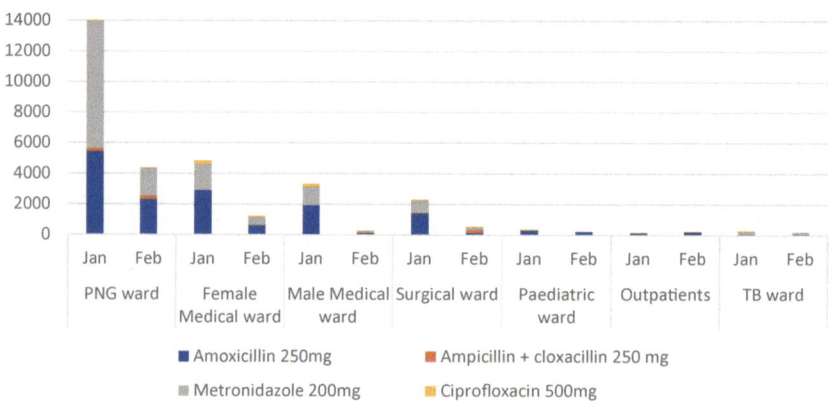

Fig. 6.1 Supply of oral antibiotics to all wards in January and February 2020

Table 6.2 Intravenous antibiotic consumption in Post-natal and Gynaecology wards at FPRRH, December 2019–March 2020

	Metronidazole 500 mg/100 ml	Ceftriaxone 1 g/vial	Meropenem 500 mg/Vial	Ampicillin 500 mg/Vial	Ciprofloxacin 200 mg/100 ml	Benzathine 2.4 MU/Vial
Standard dosage	500 mg, 3 times daily ×3	2 g, daily × 5	500 mg, 3 times daily ×5	500 mg, 4 times daily ×5	400 mg, 2 times daily ×3	Weekly ×3
December	940	699	0	3	8	1
January	2357	1262	54	12	4	0
February	1090	252	18	0	0	0
March	1348	1042	0	0	0	0
Total	5735	3255	72	15	12	1

Table 6.3 Oral antibiotic consumption in post-natal and Gynaecology wards at FPRRH, December 2019–March 2020

	Metronidazole 200 mg tabs	Amoxicillin 200 mg caps	Ampicillin & Cloxacillin 500 mg caps	Ciprofloxacin 500 mg tabs	Doxycycline 100 mg caps	Cefuroxime 500 mg tabs
Standard dosage	400 mg, 3 times daily × 5	400 mg, 3 times daily ×5	500 mg, 2 times daily ×5		100 mg, 2 times daily ×7	1000 mg, 3 times daily ×5
December	7477	2400	732	410	144	0
January	8298	5436	220	102	182	10
February	1812	2308	235	20	153	0
March	7474	4860	0	650	88	0
Total	25061	15004	1187	1182	567	10

dominance of metronidazole and amoxicillin. Both sets of data show large fluctuations in consumption over the 4-month period with particularly depressed consumption in February:

The fluctuations identified above suggest serious supply issues determining access to antibiotics. With current systems in place, the major challenge facing the MSI model is access to the right antimicrobials at the right time. The laboratory scientist is clear about this:

> Antibiotic stock-outs remain a serious constraint; in many cases patients can only be given the right antibiotics if they pay and many of them can't pay. **We have to be very clear, antibiotic stock-outs are a key factor fueling AMR**. If we look at the scenario where we have investigations done and antibiotics are available, and the outcomes are good, but we have done the investigations and the antibiotics are out of stock we won't have a good outcome.

There is a bigger concern here too; if supplies are not available and the ward staff are unable to respond effectively to laboratory results, this can be predicted to have a major impact on staff motivation and the behaviour change gains achieved. Problems of access critically restrict pharmacy's ability to engage in rational prescribing (prescribing the drug most likely to work according to laboratory results). This in turn leads to over-consumption of poorly performing (and often cheaper) antibiotics and poor patient outcomes. The pharmacy team involved in procurement planning pointed out the severe budgetary constraints they faced when attempting to order new antibiotics in response to laboratory results. In practice, this meant making difficult 'trade-offs', especially when the new antibiotics are so much more expensive than those they were able to reduce. The reduction in supply of amoxicillin, for example, is explained as follows:

> We realised that the majority of patients using amoxicillin were mothers discharged after giving birth. They are usually given amoxicillin as prophylaxis to prevent infection. Some were being given for a longer duration than necessary. As pharmacy staff, we intervened so that the duration of treatment would be reflective of the nature of risk. This led to a reduction in use. We had to increase certain antibiotics or include new antibiotics as well. Due to budget constraints, it was agreed during the planning stage that we cut on the quantity of Amoxicillin to free up some budget to cater for other needed antibiotics.

The laboratory results indicated very high levels of resistance to both amoxicillin and ampicillin, both of which are derived from penicillin. Based on these assumptions, ampicillin and amoxicillin will have minimal effects on the three primary bacterial causes of infection. Significant changes in planned use of meropenem can also be directly attributed to the MSI, where imipenem was shown to be 100% effective against *E. coli*. The pharmacist explains that consumption of this drug over the past year has been reliant entirely on donated supplies (it was not ordered in 2019):

> [The increased order] can be supported by evidence generated by the laboratory. Due to the increased culture and sensitivity reporting, we noticed that there was improved sensitivity to meropenem. This ensured that we were able to convince members involved in planning to include it on the 2020/2021 plan. We were able to get some donations last year and that's why it shows that we consumed it. What's more, we wrote to NMS to allow us to procure it, even though it's not in the current plan.

This action, of communicating directly with NMS on procurement, represents one example of where the pharmacy team have attempted to advocate as a result of the MSI. The impact of this procurement may be to the benefit of other hospitals if NMS are influenced to place it on their catalogue in future. The decision to increase orders of meropenem required the team to make stark choices which led to the reduction in orders of cefotaxime:

> Just like Meropenem, this particular consignment of Cefotaxime (used in 2019) was a donation. It wasn't in the procurement plan. While working on the 2020/2021 plan, we had to prioritise between Cefotaxime and Meropenem. We had to go with Meropenem. We did factor in the cost and resistance profile per the lab reports.

Cefotaxime has shown high levels of resistance in the laboratory tests. The marked rise in procurement of ciprofloxacin is also directly attributable to the MSI although the pharmacy team were concerned about the volume needed:

> What we require is actually a lot more. Again, [the increase] can be explained by the results of culture and sensitivity. There seems to be less resistance to ciprofloxacin.

The pharmacist sums up the impact that the project has had on procurement planning and the constraints the team had to work with:

> We had to reduce the quantity of Ceftriaxone by a significant margin. This was supported by laboratory data which showed a lot of resistance to ceftriaxone. Some of the monies freed up were used to plan for chloramphenicol and meropenem, drugs which are showing less resistance as per lab reports. There is no significant increase in the incoming budget for drugs and medical sundries. It's therefore painstaking to reallocate priorities in terms of drugs while maintaining the same budget. Our [MSI] efforts to encourage and support Culture and Sensitivity testing and sharing this with the procurement planning team lead us to include some much-needed antibiotics (Meropenem and Chloramphenicol) in next year's plan and reduce the procurement of antibiotics with a lot of resistance (Ceftriaxone).

Table 5.11 also provides an indication of the cost implications of the changes in antibiotic procurement because of the MSI. Most of the increases in procurement involve more expensive antibiotics. The procurement plan reflects the negotiations the pharmacy team have engaged into balance the need for rational prescribing against the cost implications of

Table 6.4 Order fill rate at FPRRH (2019/2020)

Financial Year Cycle	Total Items Ordered	Total Items Delivered	Fill rate (%)
CYCLE 1 (July—August 2019	307	236	77
CYCLE 2 (September—October 2019)	306	232	76
CYCLE 3 (November—December 2019)	309	226	73

Source Rx on-line medicines management system (NMS do not provide data on fill rates for specific medications)

buying more expensive antibiotics. Unfortunately, the constraints of the NMS budget-line are not the end of the story.

DISCREPANCIES BETWEEN ORDER AND SUPPLY (ORDER FILL RATES)

In practice, not all that is ordered by the hospital from NMS is supplied. The 'Order Fill Rate' gauges the delivery performance of total number of items ordered against the total number of items delivered. As clearly seen in Table 6.4, NMS supplies about 75% of orders. More specific discrepancies may also arise. Unusually, NMS failed to deliver ceftriaxone in September 2019, for example.

KEY CHALLENGES TO SUSTAINED BEHAVIOUR CHANGE: THE IMPACT OF STOCK-OUTS ON AMS

Stock-outs (the exhaustion of supplies) are a feature of Ugandan public health facilities at all levels and a major factor contributing to sepsis deaths in maternal and new-born health (Bua et al. 2015). Inevitably, bi-monthly deliveries tend to be exhausted quite rapidly and often by the end of the first month:

> For example, when we get 6000 vials of Ceftriaxone, we consume all of it in maybe 4 weeks, then we stay without for another 3-4 weeks. And the

following cycle, we get the same quantity. Therefore [consumption data] are merely an average of what is not enough.

Stock-outs are caused by a combination of misuse and overall shortfalls. The pharmacists noted that the project had improved antibiotic use on the wards:

> When these antibiotics are received [from NMS] they tend to run out quickly. Again, this is attributed to the small budget and probably misuse/irrational prescribing. However [MSI's] endeavour to link the ward, pharmacy and the Lab has to a great extent solved the issue of irrational antibiotic prescribing.

The issue of stock-outs remains a serious block to progress. An intern doctor describes the shortages affecting his work that are likely to result in Surgical Site Infections:

> We always have challenges with supplies from National Medical Stores (NMS). We have them, then for 2 months we don't have any. They deliver every quarter but in the last 6 weeks we have none – there is no oxytocin now. We have no iodine in theatre, so we are just using saline and during that time we need some help until NMS provide for us. We order for 3 months but things get used before 6 weeks. These are the challenges that make us get sepsis on the ward.

To illustrate the breadth of this shortfall, on February 18th, 2020, the PNG ward contacted K4C to request support in the purchase of gauze. Without this, they would not be able to continue with the wound dressing established on the ward. This would have resulted in increased infection and sepsis (and antibiotic consumption). We were aware during the project visit in January 2020 that the hospital had also run out of disinfectants, iodine and spinal needles (amongst many other things). In such circumstances, the only option is for staff to ask patients to pay for the necessary items, and if they are unable to pay, then operations will not happen, and major delays occur in treatment. On February 19th, 2020, we established that 13 key items for use on the PNG ward were out of stock and had been for over a month:

1. Ceftriaxone injection
2. Intravenous metronidazole

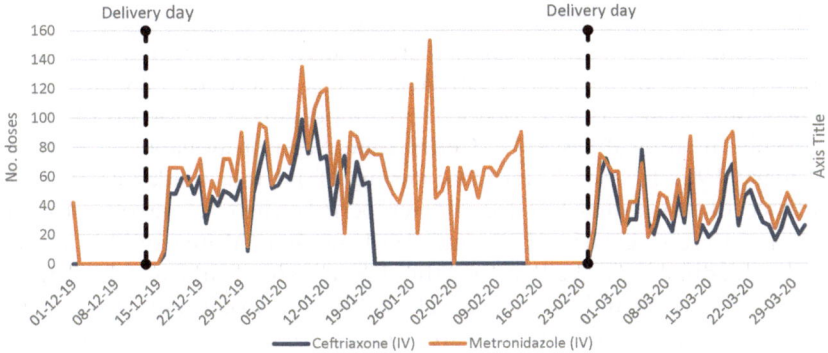

Fig. 6.2 Dispensing of IV Ceftriaxone and Metronidazole from in-patient pharmacy to PNG wards at FPRRH (1 December 2019–29 March 2020)

3. Intravenous normal saline 0.9%
4. Intravenous ringers lactate
5. Intravenous ciprofloxacin
6. Meropenem 1 g injection
7. Gentamicin 80 mg injection
8. JIK (sodium hypochlorite) solution
9. Alcohol hand gel
10. Chlorhexidine gluconate 4%
11. Cotton wool 1 kg (hospital quality)
12. Gauze (hospital quality)
13. Povidone Iodine.

The next supplies were expected on February 25th, 2020.

We have listed the items to illustrate the profound impact of such stock-outs on ward cleaning; hand hygiene[5]; the ability to clean and dress wounds and prescribe necessary antibiotics. Figures 6.1 and 6.2 present data collected manually from in-patient pharmacy on key antibiotics used in the PNG wards. NMS deliveries were made on December 13th, 2019, and February 24th, 2020. The dispensing patterns show stark evidence of stock-outs.

[5] K4C makes up this shortfall.

The greater consumption of metronidazole can be attributed to the high dosing regimen (three times daily) plus its empirical indication as a broad-spectrum therapy for prophylaxis against anaerobes. IV Ceftriaxone (dosed once daily) is also being used empirically and for prophylaxis, especially in surgical cases. IV Ceftriaxone was not available at all from February 19th to March 1st. Dispensing of IV metronidazole showed evidence of stock-outs but for shorter periods than ceftriaxone. In the period between February 15th and 23rd, neither IV metronidazole nor IV ceftriaxone were available to the PNG wards. Similar falls can be seen at Christmas and New Year. These dips we would anticipate being due to lack of doctors on the wards to prescribe (so a human resource management issue rather than a stock-out issue) (Fig. 6.3).

Dispensing patterns for oral metronidazole show higher utilisation and longer periods of stock-outs than oral amoxicillin, with an extended stock-out from January 26th to February 23rd coinciding exactly with the stock-out of IV metronidazole.

The interview with the hospital administrator raised this issue. He explains the lack of autonomy the hospital has in relation to procurement:

> [Interviewer] The biggest blockage seems to be stock outs of antibiotics, jik and disinfectant. It seems when they deliver after 1 month it is all gone so you have a month with none so even if we do the swabbing and testing if the right antibiotics are not in stock, we can't do anything. It seems you

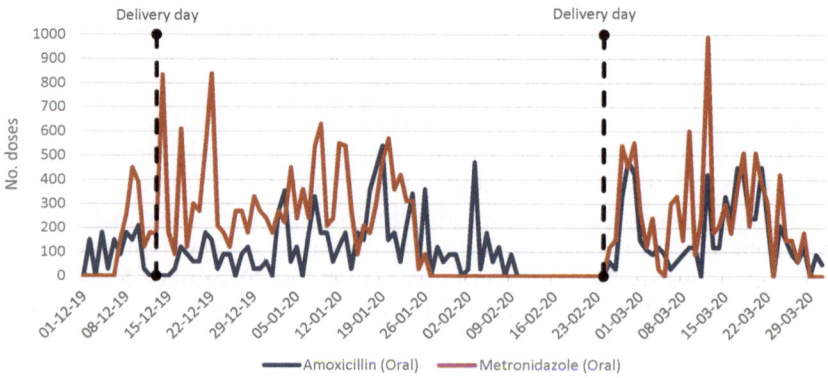

Fig. 6.3 Dispensing of oral Amoxicillin and Metronidazole from in-patient pharmacy to PNG wards at FPRRH (1 December 2019–29 March 2020)

are saving a lot of money, but we are not allowed to use this on drugs and consumables. If that doesn't improve, I don't think we can make any more improvement. Do you have any ideas?

Well NMS should be providing. The funds for buying medicine don't come from our account – it comes from NMS. We are tied to NMS.

Concerns about stock-outs were also raised in the IPC Committee. It was interesting to observe that this issue is so endemic and structural that it was almost not listed in the meeting minutes and would not have been if one of the K4C team did not raise it. Concern was also expressed by a doctor at the IPC meeting about overuse of antibiotics in some departments contributing to shortages: *'Can we stop some departments using antibiotics for everything – coughs ... malaria...'*. A senior nurse participant echoed this concern referring to the practice of prolonged antibiotic use: *'clinicians prescribe antibiotics for 10 days then the clinician just says 'continue'*. This is one of the practices that we have noticed to have changed on PNG since culture and sensitivity testing became part of the ward's normal practice.

The K4C pharmacist responded to these concerns by expressing the need for the full institution of a Medicines Therapeutic Committee (MTC) which would have authority to implement clearer policies on antibiotic use.

Interpreting In-Patient Pharmacy Data—A Note

Consumption data cannot be interpreted in any direct way as an indication of prescribing knowledge or stewardship competency. Collecting and analysing the data on antibiotic 'consumption' patterns at FPRRH have emphasised the dangers of empiricist approaches to data analysis and presentation and the importance of interrogating data rigorously. In most cases, data presents a myriad of questions and very few obvious and immediate answers. This is especially the case when attempting to collate data from public health facilities in Uganda and many other LMICs. Collecting data has been a continuous process of trial and error merging with the underlying ethnographic journey. As described, in the case of in-patient pharmacy, the data is not yet managed electronically via the Rx system. Rather, individual wards come to a window in the in-patient pharmacy with paper forms requesting supplies for that day. The pharmacy then maintains a handwritten record of dispensing in a records book, and this data is compiled into forms for the Ministry of Health. In the first instance, we believed that all wards behaved in this manner, but the initial data suggested otherwise. The TB ward, for example, appeared to collect very few drugs. We found that most of the drugs used on the TB ward are in fact provided by a donor[6] and follow a different track. These drugs pass through the main stores and are directly requisitioned by the TB ward. As such, they do not pass through in-patient pharmacy and neither are they recorded as received from NMS on the Rx system, leaving a gap in overall supply and consumption data.

Other apparent 'discrepancies' in the data, including very sporadic and low use of antibiotics in the neonatal intensive care unit (NICU) and the paediatric ward, uncovered variances in practice that are undocumented and apparently do not comply with the published protocol. This was explained by one respondent as follows:

> The paediatric ward gets injectable drugs (directly) from main stores, including antibiotics, but oral antibiotics and other oral drugs from in-patient pharmacy.

Our observational work on the wards supplemented by qualitative interviews and many emails led us to question the relationship between this 'consumption' data and overall consumption patterns. We know, for example, that since laboratory results have been available, many women are getting higher-end antibiotics and that this is contributing to shortened stays and improved patient outcomes. However, the supplies of these drugs were not visible in the data. It seems that the pharmacists have played a critical role in supporting access to these antibiotics through a combination of 'borrowing' from other hospital supplies. The following excerpt explains this process:

> There is a TB focal person who handles all TB related logistics, including the ordering of TB drugs. These drugs are stored in main stores. Rx only focuses on drugs from NMS. In most cases, we don't have the changed antibiotics in stock or in our procurement plan or we have limited quantities. Take Amikacin for example. We usually borrow from the TB

---

[6] The Ministry of Health's National Medicines Policy reports that donor support for medicine financing is more than three times the amount the government invests and concludes that, 'heavy dependence on donor funds puts Uganda in a vulnerable position (2015: 7)'.

program and give to the patients. The same applies to Moxifloxacin. These medicines are not available in the inpatient pharmacy. In fact, one time you had to use K4C money to purchase Amikacin. Because at times we have septic patients who are only responsive to these drugs, we 'beg'/borrow from the TB drugs. Apparently, this has caused audit queries.

This 'borrowing' behaviour clearly saves women's lives; it also compromises the pharmacists under pressure to assist, but potentially contravening donor conditionalities. We can anticipate similar situations in relation to Anti-Retroviral Therapies (for HIV patients) and, in the case of FPRRH given its proximity to Congo, some (necessary) stockpiling in the event of Ebola spread.

Pharmacy Support in Accessing and Distributing Antimicrobials

It is important to recognise the considerable efforts the pharmacy team have made to optimise the distribution of and access to supplies within the hospital. Improved relationships with pharmacy have significantly improved the ability of staff in the PNG ward to access consumables and medicines. The hospital pharmacist explains his role and how he has tried to improve timely access:

I was very happy to be part of the project; most of the things I was trying to do beforehand but I was focused in stores on the supply chain. This is a key role for hospital pharmacists in Uganda. Making sure the necessary consumables are available; that people are using standardised protocols for theatre, but they often don't have consumables. I have a level of authority to ensure consumables are available; to develop good pharmacy distribution practice. The hospital has supplies for different programmes (ART, TB and emerging diseases such as Ebola). I can help to move items around to ensure user access and good distribution. I can ensure different wards have access to medicine so they can do their job. I am involved in proper procurement planning for the entire financial year; how do we move drugs to the user points? I think the commodities for sepsis need to be ring fenced somehow. Some come under different programs; we use one antibiotic that comes here under TB, so we redirected it, and this has been very beneficial.

Table 6.5   Patients receiving the prescribed antibiotic dose

	Gynaecology (n = 22)	Post-natal (n = 20)
No. of patients on antibiotics	10	18
No. of patients documented to have received all prescribed doses in last 24 hours	4	0

Source Results of G-PPS, May 2019 as reported to FPRRH IPC Committee

The midwives have appreciated the lengths pharmacy have gone to try to access antibiotics and consumables:

> The relationships with pharmacy are working a lot better; they get us wound dressing materials and [pharmacist] runs to the ward when he is needed.
>
> There was a case who is only sensitive to amikacin. Getting one dose of amikacin is very expensive and our population here cannot afford. It isn't available in the hospital; they can't afford to provide it sustainably as it is very expensive but now [pharmacist] tried hard and we have been able to provide, and she is greatly improved.

When drugs are prescribed but not in stock, patients are asked to buy them privately.[7] Current recording systems cannot capture the consumption of privately purchased drugs. By way of illustration, the GPPS 'snapshot' found that, on the date of the survey, very few of the patients were receiving the prescribed dose (Table 6.5):

These examples underline the need to exercise caution when interpreting data on antibiotic consumption. We have noted how the presence of laboratory results has facilitated clinical pharmacy on the PNG ward. The presence of one clinical pharmacist has made a major difference. Extending this to other wards may, however, require more pharmacists to be employed.

---

[7] The regulation of 'over-the-counter availability and self–medication with antimicrobial medicines' is a key stewardship target of the National Action Plan. We were very aware of this huge challenge in Uganda and are undertaking some work to assess the quality of some of these antibiotics. We made the decision not to address this issue in the MSI. Documentation on the ward was also so weak we decided not to attempt to capture the role that private purchases play in antibiotic consumption on the ward. This is an area we would be keen to follow up in future.

Pharmacy Leadership in a Ugandan
Regional Referral Hospital

Chapter 1 noted the emphasis on pharmacy as a profession in the Call for Applications and, in the 'Scoping Requirements', a preference for pharmacy leadership of AMS. In order to understand the current and potential role that pharmacy, as a profession, can play in the management of AMR, it is important to understand the constraints on pharmacy in context. The role of a hospital pharmacist in a Regional Referral Hospital combines supply chain management with clinical engagement and includes (in theory):

- Monitoring the supply of all medicines used in the hospital, purchasing, compounding, dispensing and quality-testing medication stock.
- Working with the healthcare team to ensure the selection of the best medication at the correct dose for an appropriate duration.
- Monitoring and preventing or minimising side effects and drug interactions.
- Providing medication counselling to patients.
- Dispensing medications for patients in wards, the emergency department and those attending outpatient clinics.
- Offering specialist drug therapy advice to doctors, nurses and other healthcare professionals within the hospital.

When the MSI commenced, there was only one senior pharmacist in the hospital. The Head of Pharmacy in a Regional Referral Hospital also coordinates pharmacy-related activities across the wider region. Fort Portal Regional Referral Hospital is the referral hospital for the Rwenzori region comprising 9 health districts and an approximate population of 3 million people (2014 National Census).[8]

The hospital also employs a pharmacist, who reports to the senior pharmacist. Unfortunately, when the project commenced, the pharmacist had been transferred to the National Referral Hospital. In addition, the hospital hosts intern pharmacists allocated by the National Internship Committee under the Ministry of Health. Internships last 12 months of

[8] https://www.ubos.org/2014-census/. The National Medicines Policy reports that only 8% of pharmacy posts in public service are filled (MOH 2015: 10).

which 6 months must be spent at a hospital with rotations to different sectors including regulatory bodies, community pharmacies and pharmaceutical industries. The hospital usually has four intern pharmacists who rotate through the major wards, pharmacies and stores under the supervision of the pharmacist/senior pharmacist. The hospital also employs two Pharmacy Technicians and two Pharmacy Assistants who are qualified to diploma and certificate level, respectively, and are mainly involved in dispensing. A second pharmacist joined the pharmacy team during the project and was required in the first year of his appointment to develop systems to improve supply chain management. According to the Ministry of Health's[9] plan, there is a vacant position at FPRRH for a principal pharmacist. In this context, there is limited pharmacy capacity to engage more actively in clinical pharmacy across the whole hospital.

This chapter has outlined the opportunities that more active engagement with the laboratory has brought about particularly in stimulating the practice of clinical pharmacy on the wards. It illustrates the gains that can be made from the more effective deployment of just one pharmacist. It has also identified key structural weaknesses in the health system that restrict the opportunity for optimal impact and scale-up. Chapter 7 explores the change process in more detail and the contribution that multi-disciplinary team working has made to knowledge mobilisation and change processes on the ward.

REFERENCES

Ackers, H. L., Ioannou, E., & Ackers-Johnson, J. (2016). The impact of delays on maternal and neonatal outcomes in Ugandan public health facilities: The role of absenteeism. *Health Policy and Planning, 31,* 1152–1161.

Baine, S. O., Kasangaki, A., & Baine, E. M. M. (2018). Task shifting in health service delivery from a decision and policy makers' perspective: A case of Uganda. *Human Resources for Health, 16,* 20. https://doi.org/10.1186/s12 960-018-0282-z.

Brink, A. J., van de Bergh, D., Mendelson, M., & Richards, G. A. (2016). Passing the baton to pharmacists and nurses: New models of antibiotic

---

[9] The MOH allocates staff to Regional Referral Hospitals centrally; hospitals have little autonomy in recruitment.

stewardship for South Africa, *South African Medical Journal*, *106*(10), 947–948.

Bua, J., Mukanga D., Lwanga M., & Nabiwemba, E. (2015). Risk factors and practices contributing to newborn sepsis in a rural district of Eastern Uganda: A cross sectional study. *BMC Res Notes*, 8, 339. Published online 2015, August 9. https://doi.org/10.1186/s13104-015-1308-4.

Ministry of Health, Uganda. (2015). *Uganda National Medicines Policy.*

Tweheyo, R., Reed, C., Campbell, S., et al. (2019). 'I have no love for such people, because they leave us to suffer': A qualitative study of health workers' responses and institutional adaptations to absenteeism in rural Uganda. *BMJ Global Health*, 4, e001376. https://doi.org/10.1136/bmjgh-2018-001376.

World Health Organisation. (2007). *Task shifting to tackle health worker shortages*. Retrieved from: http://www.who.int/healthsystems/task_shifting_booklet.pdf. Accessed 4 March 2017.

World Health Organisation. (2008). *Task shifting. Global recommendations and guidelines*. Geneva: World Health Organisation.

World Health Organisation. (2012). *Optimizing health worker roles to improve access to key maternal and new-born health interventions through task shifting*. Retrieved from: http://apps.who.int/iris/bitstream/handle/10665/77764/9789241504843_eng.pdf;jsessionid=876E08C843C919EA6CE890EB370BEAB8?sequence=1. Accessed 1 February 2018.

CHAPTER 7

Change Processes: Multi-Disciplinary Teamwork

Abstract This chapter examines the importance of multi-disciplinary team-working to the management of antimicrobial stewardship and behaviour change processes. MDT has played a major role in improving communication and reducing professional boundaries that had previously contributed to high levels of infection; extended use of, often inappropriate antibiotics and poor patient outcomes. The functioning of such teams facilitates the role of clinical pharmacy.

Keywords Multi-disciplinary team working · Behaviour change · Huddling

> There is a meeting of minds of all the different people so the future for the work is quite bright. (Hospital Administrator)

> The main thing is the excellent team working you have done here. (Intern Doctor)

One of the anticipated outcomes of the project from the funders' perspective was 'improved knowledge and practice'. Knowledge comes in many forms and it is often associated too narrowly with forms of explicit scientific or clinical knowledge. The interpretation of laboratory results is an example of this kind of knowledge. Changing behaviour requires more

© The Author(s) 2020
L. Ackers et al., *Anti-Microbial Resistance in Global Perspective*,
https://doi.org/10.1007/978-3-030-62662-4_7

than an increase in this form of knowledge. Explicit knowledge works hand-in-hand with forms of tacit knowledge and it is these knowledge combinations that stimulate change. Our research on the gains to NHS staff from periods of volunteering in Uganda (Ackers et al. 2017) has emphasised the relative importance of tacit knowledge; of experience in leadership, cultural understanding, communication, and team working. Team-working, as much as any gain in scientific or clinical know-how is the connecting knowledge that has made this project a success. Indeed, it is the positive experience of team-working that has driven the quest for scientific knowledge (to know more). This fits with our conceptualisation of behaviour change; it is not new knowledge that drives motivation; but motivation that drives the thirst to learn. It is interesting to reflect here on the assumed knowledge gains to NHS volunteers expressed in the funding call. The emphasis on these more tacit forms of knowledge and leadership skills for NHS staff is very evident whilst the knowledge gains to LMIC staff appear to be focused on more explicit skills (prescribing and IPC). Somekh makes an interesting point about the relationship between 'training' and education or knowledge suggesting that the concept of knowledge has a more sophisticated tone whilst 'training has strong cultural associations with the teaching of low-level skills' (2006: 19). There may be other ways of understanding this relationship perhaps seeing training as more linked to technical know-how and an instruction on what to do in certain circumstances rather than a more holistic understanding of the knowledge underlying that.

It is resoundingly clear from the interviews, attendance at meetings and from on-going observational work on the ground that the key success in this intervention hinges on the achievement of genuine, multi-disciplinary team-working with each cadre showing mutual respect and willingness to learn from each other. This level of team-working did not exist prior to the project and many of the key actors had weak relationships with each other. They may have known about the existence of the laboratory, for example, but not developed a relationship with the staff there or understood the kind of contribution the laboratory could make to their work and their patients. Similarly, they may know that there are pharmacists at the hospital but did not understand their roles and knowledge and interact with them in any meaningful way, particularly in clinical decision-making. There was certainly minimal engagement with the nursing and midwifery team who are the most present on the ward and have most direct patient engagement.

The 'magic' that fostered this team working (as one respondent referred to it) is complex. Many respondents attributed it directly to the project. However, important structural changes took place in the hospital at around the same time. A new in-charge had been appointed on the ward, for example. The (externally funded) laboratory had begun to function at a much higher level. The decision to appoint an inventory manager in pharmacy stores also released a senior pharmacist with a passion for clinical pharmacy enabling him to engage directly with the team. This is an excellent example of effective task-shifting. From a project perspective, the placement of K4C staff and professional volunteers on the ward on a continuous basis undoubtedly stimulated team-working and specifically the empowerment of the nursing and midwifery team. The involvement of a senior obstetrician volunteer also stimulated the setting up of daily ward meetings which staff reported to have had an impact of team working.

The decision to improve the ward infrastructure and basic infection control; decorate; provide trolleys to enable the nursing and midwifery staff to work effectively; and the provision of instruments in the evacuation room have made staff feel valued, improving motivation and job satisfaction enabling them to use the skills they have more effectively. All of these factors are of great importance. However, we must not underestimate the impact of the introduction and embedding of culture and sensitivity testing to the team-building process. The culture and sensitivity testing process is inherently a team-based activity; it engages all staff irrespective of cadres. Having the laboratory results provides an 'objective' focus for interest to coalesce around; it stimulates team discussions and active pharmacy engagement and creates the environment for genuinely patient-centred care. Having the evidence-base for rational prescribing undermines any artificial disciplinary hierarchies (who knows best). One respondent refers to the impact of having laboratory results on hierarchies:

> Sometimes there can be ego – that the doctor or pharmacist thinks, 'I am the overall boss so I can't be directed on what to do' but with the data that goes down.

Laboratory results provide a critical focal point for what some pharmacy teams in the UK refer to as 'huddling'.[1] Gardner et al. characterise huddling as a mechanism to 'support inter-professional communication and collaborative practice' (2018: 16). The consequences of rational prescribing as a central component of patient care have delivered tangible and often quite surprising results. Staff all remember and refer directly to cases where women's lives have been saved and women have got better remarkably quickly as a result of this process. The sheer tangibility of these outcomes has had a major motivational effect on all concerned; so, team-working becomes enjoyable and immediately rewarding. We cited the words of a Ugandan health worker in our previous book to illustrate the devastating impact on health worker motivation of having 'people dying in your hands [when] you know what to do but don't have the things to help them' (Ackers and Ackers-Johnson 2016: 96). Seeing patients survive and leave hospital has a huge motivational impact and drives a thirst for knowledge. Some of the potential rewards to working on the PNG wards relates to the nature of the work itself. Many respondents spoke of the negative aspects of the work before the project in terms of the offensive smelling wounds. Once this problem was resolved and women were seen to be getting better quicker, health workers noted the rewards of working in this area:

> The [nurses and midwives] are proud; they really feel good when they say they don't have any septic patient on the ward which means they are trying to prevent sepsis.

When we asked a midwife if she had had experience of this kind of team working in other areas, she replied that team working did coalesce around obstetric emergencies on labour ward. But she made the interesting point that these emergencies are short-lived events and the fact that women with sepsis stay in the hospital for longer helps to build these relationships with patients and a more vested interest in, and responsibility for, their well-being:

> [Have you seen that teamwork, motivation, and collective interest in the patient in any other place you have worked?]

---

[1] http://www.pulsetoday.co.uk/pulse-intelligence-your-practice/regulation/how-our-morning-huddles-improved-practice-teamworking/20037047.article.

In labour suite when there is an emergency there is teamwork but once the emergency is over, they move on. But here the patients stay longer so we see the same patient. We want to know how they are doing – if they are getting better so there is continuity of care here.

A knock-on consequence of this has been the fact that the ward is no longer 'the laughingstock' of the hospital; and is seen rather as a centre of excellence to be envied. This has been openly acknowledged in hospital meetings as having a massively motivational impact on all staff concerned. The in-charge describes this change:

Things have totally changed. Before the project, the ward faced challenges. The hospital assesses every unit regularly along various indicators. Before, the unit always came out badly and we were always hearing our name in meetings as a result; but now things are much improved. The hospital director said to me; 'What's happening on post-natal? You people are quiet - the deaths are not there as they used to be'.

The Chair of IPC Committee echoes this view:

Sincerely the trend has changed. Maternity was the ward we laughed at and now it is the ward we all envy.

When a team is successful, and its success is acknowledged, new members seek to join the team and gain that experience. And as they leave the ward, they are motivated to take that experience to their new ward or facility. The following section explores some of the team working relationships in more detail and identifies areas for future improvement.

PERCEPTIONS OF TEAM WORKING

Every respondent reported the improvement in teamwork and identified this as the source of change on the wards.

We are now working hand-in-hand with the pharmacists, the laboratory, and the doctors'. (In-charge)

I saw a gap with the midwives and doctors. You pass by on a ward round they don't always tell you what challenges they face. I found a burst abdomen and the pus was a lot, so I found myself cleaning the wound.

> On ward rounds you don't usually see the wounds so you can't know the problem. Now I combine the ward round with wound dressing (it can take me up to 10 pm). We do the wound dressing together with the midwives, so we work together. (Intern Doctor)

One midwife describes the team-working that has developed. She particularly emphasises the engagement of doctors which she puts down, in part, to the sheer tenacity of the K4C midwives which they have now 'inherited':

> We are working hand-in-hand together as a team. Before the doctor was not there but nowadays doctors are responding. (The change) is because of the teamwork those people introduced. They showed us good leadership. They would follow-up and afterwards they call the doctor and he comes even if they first deny we call again even the senior doctors. They come – it is a big change.

The reference here to senior doctors is interesting. The general view reinforced by observations on the ward and the experience of a strike by intern doctors, in part due to lack of supervision (in September 2019), is that interns are largely left to cope and the presence of senior doctors is at best sporadic. The relatively junior status of the intern doctors may, to some extent, explain their willingness to engage in team-working, benefit from mentoring by the pharmacist and their appetite for learning. However, concern was also expressed about intern 'burnout' in the absence of supervision and the challenges of continuity when interns rotate every 2 months. The professional volunteers play an important role in supporting interns in the absence of senior Ugandan doctors. The in-charge nurse took on a major role in orienting new interns on the ward which will help them to integrate in the team but this puts an additional burden on the in-charge and could and should be fulfilled by more senior doctor engagement. It was clear from the discussion in one of the IPC Committees that the tension between doctors persists in other areas of the hospital and was previously evident on the PNG ward:

> Sometimes you find a pharmacist has changed a prescription and then on the ward round the doctor changes it back to a drug the patient is resistant to (Midwife on IPC Committee)

A pharmacy intern reflects on this situation commenting:

I think it should be teamwork here and respect for each other. If we advise and then the prescription is changed the clinicians come and undermine that decision without listening to the pharmacist. We can see that post-natal is taking the lead in consulting with pharmacy, but other units are relying on empirical usage designed for health centres and not for hospitals. If you are rigid on the usage you will not use the pharmacists/laboratory's advice.

Another midwife is appreciative of the team-working environment that has developed. She refers to the presence of pharmacy on the ward and the lengths the pharmacists have also gone to, to try to secure appropriate antibiotics:

It has greatly improved because right now we have pharmacists who come on a daily basis or if not, every day we can't go 3 days without seeing one who can guide us on the mothers and which drugs to take. They interact with the doctors; if you don't interact there is that collision. Right now, there is no tension. They say, 'what if we do this and there is a discussion.' We never used to have any pharmacists coming on the wards, so it was majorly the doctors dealing with the prescriptions. We have managed to reduce the irrational use of antibiotics.

The impact of medicine on multi-disciplinary team working is well established internationally. Atwal and Caldwell (2006) identified three barriers to effective team working in acute care (in the UK): these included, the 'dominance of medical power that influence interaction in teams' ((2006): 359). In a more recent editorial with specific reference to models of antimicrobial resistance in South Africa, Brink et al. consider whether models of antimicrobial stewardship developed in high-income countries translate effectively to a South African public health care context. They underline the lack of understanding of the contextual and behavioural determinants that influence local prescribing practices and present alternative models using non-specialist pharmacists, nurses and even community health workers in key stewardship roles. As doctors themselves they pose the question; '*Will doctors, the actual prescribers, and a group notoriously keen on safeguarding clinicians prescribing autonomy, embrace pharmacists and nurses as equals in the drive towards optimising antibiotic use?*' (2016: 947).

The following midwife makes a similar observation and illustrates the point above about the motivational effects of seeing a patient's condition turned around:

> Now [the pharmacist] comes in daily and helps us and discusses cases with us. We had a discussion with him on Friday. He comes and looks at the files and the wound; the drugs; the length of days the patient has taken and then he gives us some alternatives. He looks at the laboratory results. There was a time he prescribed a very expensive drug and the healing was very fast. We nearly lost that young girl of 16. The baby died. They took her to theatre twice as her uterus was necrotic, but the moment we brought that drug she improved. She has her life.

An intern doctor describes how useful he has found the specific expertise of the pharmacy team on the ward:

> It's changed a lot now; the senior pharmacist comes regularly, and you may find there are 2 microorganisms sensitive to different antibiotics. Now I don't have the time to walk to the pharmacy and those people have studied medicines. I have textbook knowledge; if someone has a UTI (Urinary tract infection) I give x. I did study this but as time goes on you get used to giving certain medicines quite often and you are not so equipped to understand how one medicine interacts with another one or if a patient has TB or is HIV positive or how to combine drugs – so a pharmacist being available on the ward has really brought great improvement.

Evidence suggests that hierarchical structures inhibit junior staff from challenging their seniors, leaving them to prescribe in a way to suit conventional senior preference rather than offering evidence-based treatment options (Broom et al. 2014). The MSI has demonstrated that effective team working challenges professional hierarchies and empowers lower cadres of staff. When team working is based on collegiality, recognition of respective skills and the role they can play in helping (especially doctors) to achieve their goals efficiently can drive change. Another senior doctor (in the IPC Committee) referred to the pharmacists as a 'tie-breakers'.

'We Are a Bridge': The Changing Role of Pharmacy in FPRRH

The intern pharmacist who later came to be employed on the project describes his experience of the role of pharmacy and how that has changed since the project commenced:

> Interns shadow the pharmacists and I shadowed (senior pharmacist) to gain experience under supervision. Unfortunately, the pharmacists' role (in an RRH) is limited to the supply chain. One of the challenges in a hospital job description – we know we can do a lot but when you get employment in a hospital there isn't scope for clinical pharmacy. We try as much as possible to show what we can do on a clinical side and we want to specialise. You need to specialise if you want your work to be noticed; at present people just see us moving around; you are everywhere, and your work isn't noticed. There should be different roles for pharmacists within the hospital and scope for clinical pharmacy. On the post-natal ward, for example, there are 6 + doctors and 1 pharmacist (shared with the rest of the hospital). It's not effective.
>
> [But that means employing more pharmacists?]
>
> We – and the Pharmaceutical Society of Uganda (PSU) are advocating for specialisation. At present we manage the entire supply chain from planning, procurement, monitoring use, distribution, some physical inspection (are they keeping medicines in the right condition) and antibiotic use. We are supposed to review prescribing in the in-patient pharmacy; was it given for the right condition? And we should go through the notes, but we don't really have time to do 40 files at once. We can't review so we are just forced to dispense. So, you end up dispensing which is one of the key challenges. We want to do more, but you are really limited. It's frustrating for my future career – there is no direct career path in Uganda. The PSU came up with a draft career plan for pharmacists to discuss with the Ministry of Health. The government has been giving scholarships for clinical pharmacy but there are no positions for such people in public or private hospitals. So, why do a masters if I won't get employment anyway?

This concern about career paths is echoed in many other professions in LMICs. Creating new opportunities for higher-level qualifications and professionalisation—often encouraged by foreign experts seeking to emulate developments in their own systems (the benefits of which are often contested)—may have serious unintended consequences when higher level and remunerated positions do not evolve in parallel. This has been the case with degree-level programmes in nursing, midwifery and biomedical engineering, for example. There are clear benefits to higher-level capacity-building programmes when appropriate positions are available. Failing to provide such positions compounds frustration and is highly demotivating as graduates return to the same badly paid positions. There are other consequences too as graduates grasp the opportunity to avoid clinical work and seek out work with foreign NGOs. Expanding clinical pharmacy in Uganda requires both higher-level qualifications; the development of appropriate positions AND a significant increase in the volume of pharmacists in the hospitals to release them from supply chain management roles.

He then talks about his own experience of working on the project and the mentoring he has received from the pharmacist:

> I'd like to do clinical pharmacy. The pharmacist has lots of experience and he is biased towards the clinical side. He is a good mentor, but he has the same frustration. He has made me appreciate the value of pharmacists in providing clinical services.
>
> The project has changed my role a lot. There is an opportunity to really delve into what AMR is. We had heard of IPC, but we had not participated in any. I studied the theory of it at university, but this has made me learn a lot about culture and sensitivity testing and how to interpret the results. There is a key role for us (pharmacists) to interpret and guide clinicians – I've really learnt a lot! At university it was all theoretical. Now we are learning about culture and sensitivity testing and how to interpret the laboratory results. (The pharmacist) was able to show me some of things so that is an area I gained a lot. This is the role of the pharmacist to interpret the results and guide selection of antibiotics.
>
> Pharmacy were not really doing that role before. Their job description was limited only to supply chain management. This has changed and people are now very interested – and this is because of our project. We wanted to do these things, but the project created the opportunity for us to get involved. We only became involved because of the project; the clinicians were not so keen before to involve the pharmacists – it wasn't conflict as such, but they didn't appreciate the role that pharmacy brings. At school we study together but there are no tensions as such now (with the intern doctors) now they know what pharmacists can do.
>
> The clashes come when you try to perform their roles. For us we don't do diagnosis. You should each stick to your sides, but they recognise on the issues of antibiotic selection, they need our guidance. **We are a bridge** and we also know what is available in the stores and they appreciate that.

We have information about if antibiotics are available and where we can get them from. They really appreciate it, especially the younger doctors.

His ability to extend himself and engage in clinical pharmacy reflects his supernumerary status. When we decided to take steps to retain this pharmacist at the end of his internship and support his salary through the project, we developed a clear role description so that he did not just become another pair of hands or substitute labour. This created a window of opportunity to focus on often neglected aspects of the pharmacy role and to demonstrate the value and cost-effectiveness of such roles in a regional referral hospital setting. He describes the importance of relationships and rapport to team-working:

As for the laboratory, well I know [Lab scientists] now. It is more about getting to know people who work there and building a rapport. Beforehand our only previous relationship was to requisition consumables from the stores for them but now our relationships with the nurses and the laboratory – because nurses take and collect the samples so there is much more rapport. The midwives have really appreciated the project – there is lots of change – they now attend changeover meetings of interns (when the interns rotate). [They] have stressed the need for interaction, so the nurses relate to what their colleagues are doing.

Asked to describe what his role now involves on the ward the K4C pharmacist explains how he and the hospital pharmacist engage on the ward:

[The pharmacist] has a passion for clinical pharmacy – he prefers to interact with patients. Until this project there was no other opportunity for him to engage in this way. He identifies sepsis cases where pharmacy can have a real input and is working especially with the intern doctors. This was a chance for him, and he is actively involved. We don't see all the cases. We try to pick cases where we feel pharmacists can have an input and that way, we don't clash so much with the doctors. We work within our mandates so there are no clashes. Intern doctors are also using this as a chance to learn about AMR.

The hospital pharmacist echoes these concerns about hospital pharmacy in Regional Referral Hospitals. He had previously worked in the national referral hospital in Kampala and had a more clinical role in that more specialised context. He referred to some resistance, in the past, from doctors to clinical pharmacy:

> In the past people didn't look at pharmacists as people involved in patient care. Doctors previously felt that undermined their authority. Some even hid the medical files from pharmacists. They tried to make us complete separate forms, so we were not working in the patient files.[2]

Asked to describe his role in FPRRH, he explains how his first year in the hospital was focused on supply chain management:

> I now come in for septic and complex cases. I go on the ward and look at the entire patient file. I look at the notes, the plan of action, see if they have culture and sensitivity testing. I even go and see the patient and look at the timeline since their admission. I started doing that when I was moved from the Stores in October 2019. I worked there for a full year which was a good way to get to know the hospital but then they recruited an inventory manager. Before he came, I was tasked with management of stores and this enabled me to leave the stores and get more involved on the wards. The new role is very challenging and very rewarding. You don't go there because you have all the answers. I have had to keep reading. I know what information and online sources to look for and where to look so we can focus on a specific case and I know how to make that information relevant to a specific case but in the process I also had to teach people; nurses and midwives. It has been a continuous conversation. The patient belongs not just to me but to all of us so we have to engage in a conversation about the patient and what we can do about it. I have found this work very motivating. I spend more time on the wards now than before and every time they have a case, they want me to look at as a specialist there is now more time on the wards for me. It has been very encouraging for me and the doctors are appreciating my presence too. In the past traditionally most resistance to change came from the doctors but they are more collegial now. They have attached a certain respect to my presence because we do this in a collegial manner. The different disciplines are coming together for a common objective with mutual respect and trust. I don't come here to criticise or undermine. In fact, when I write in the notes, I write it not as a prescription but as a recommendation. They

---

[2] It was interesting to observe during a visit with this Ugandan pharmacist during a fellowship in the UK we were shown how pharmacists used a different coloured ink to make their notes on patient files.

could disregard it. Then one doctor said, 'when he comes, and we follow his advice things turn out very well.' People's perception of pharmacy has changed tremendously. They said to me, 'we have never seen a pharmacist do what you do – how do you know to do that' and I explained to them that clinical pharmacy is part of our training. For her it was a shock; the perception was that pharmacists handle the supply chain and dispenses. I am sure the Director has had her eyes opened to what pharmacists can achieve. I think she had the same perception of every other person that pharmacists only manage procurement distribution and dispensing but now she has seen what 1 pharmacist can achieve on a very sensitive ward. The maternal mortality rate was very high and not one death has taken place since the project started. This is in line with the Sustainable Development Goals. They noted that since you guys started coming as part of our project you haven't had a maternal death from sepsis, and it used to be quite common.

Some concern was expressed at the IPC meeting, perhaps reflecting practice on other wards and not PNG that doctors were continuing to ignore the advice of the pharmacy team:

The problem was that some doctors changed the [pharmacy] recommendations and then the patient's condition did not improve so [the nurses] complained to the director in a meeting and she was very appreciative of the progress we have made.

A senior doctor challenged this attitude:

At the end of the day all our efforts should focus on the patient. The best outcome regardless of your cadre. If everyone plays his or her own role it will really be good teamwork. Without that we can't do magic here so there needs to be comprehensive teamwork and leadership. But there are issues for the administration and supplies and IPC cuts across this. Can we have more IPC supplies and more leadership involvement?

The respondent raises an important point here echoing the behaviour change literature. The development of team-working cannot happen in a vacuum and if the outcomes of that team-working, in terms of wound care or prescribing decisions cannot be operationalised due to lack of supplies then team-working will become frustrating, demotivating and collapse.

Team-working lies at the heart of the behaviour change that we have witnessed on the post-natal and gynae wards. Recognition of the contribution of colleagues from different disciplines—and at lower pay grades—has generated 'huddles' of activity that have proved highly motivational and empowering and driven a quest for more knowledge. It is important to point out that this change has happened incredibly quickly and with relatively little investment and has proved highly cost-effective. Sam Opio notes the benefits during a follow-up visit:

> There is great progress. It is clear that the departments outside of maternity are aware of the work. There is great teamwork. I was aware last time I came that people were working in silos, the laboratory, pharmacy, and the nursing team. Everything was in silos. I've seen significant teamworking. The pharmacy profile is much bigger; there is more engagement with the laboratory and the nursing team, and it is important to build on that. After documenting and disseminating we need to institutionalise and move out of silos.

Chapter 8 moves on to consider the knowledge generation and mobilisation processes that multi-disciplinary team working has enabled within the MSI.

References

Ackers, H. L., & Ackers-Johnson, J. (2016). *Mobile professional voluntarism and international development: Killing me softly?* Palgrave PIVOT. http://link.springer.com/book/10.1057%2F978-1-137-55833-6.

Ackers, H. L., Ackers-Johnson, J., Tyler, N., & A. Chatwin, J. (2017). Healthcare, frugal innovation, and professional voluntarism: A cost-benefit analysis. Cham: Palgrave PIVOT.

Atwal, A., & Caldwell, K. (2006). Nurse's perceptions of multidisciplinary teamwork in acute health care. *International Journal of Nursing Practice, 12*(6), 359–365.

Brink, A. J., van de Bergh, D., Mendelson, M., & Richards, G. A. (2016). Passing the baton to pharmacists and nurses: New models of antibiotic stewardship for South Africa. *South African Medical Journal, 106*(10), 947–948.

Broom, A., Broom, J., & Kirby, E. (2014). Cultures of resistance? A Bourdieusian analysis of doctors' antibiotic prescribing. *Social Science and Medicine, 110*, 81–88.

Gardner, A. L., Shunk, R., Dulay, M., Strewler, A., & O'Brien, B. (2018, September). Huddling for high performance teams. *Federal Practitioner*, 16–22 https://www.ncbi.nlm.nih.gov/pmc/articles/PMC6366795/.

Somekh, B. (2006). *Action research: A methodology for change and development*. Maidenhead: Open University Press.

The Knowledge Creation and Transfer Mechanism

Abstract This chapter reflects on the relationship between the knowledge mobilisation processes that have contributed to behaviour change at an individual and organisational level. It critiques the traditional emphasis in international development on one-off, formal, foreign-led 'training' episodes and contrasts these with the more fluid, bilateral, approach to learning through co-working and mentoring.

Keywords Knowledge creation · Knowledge mobilisation · Continuing medical education · Co-working · Co-presence · Mentoring

Chapter 1 noted the quite prescriptive approach taken, by the funding partners, to knowledge mobilisation processes and associated log-frame audit. This chapter examines some of the underlying mechanisms that have shaped behaviour change in the MSI intervention. Knowledge for Change has been working to support health systems change in Uganda for a decade. This has involved huge organisational learning which we often describe as learning from failure. Rajkotia similarly argues that, 'the fear of failure, instilled by the success cartel is one of the key reasons for why there is so little innovation in the health development sector arguing that 'intelligent failures arising from experimentation and exploration are 'praiseworthy' (2018: 2). Our first book, 'Killing me Softly' (Ackers and Ackers-Johnson 2016) describes the challenges involved in

© The Author(s) 2020

L. Ackers et al., *Anti-Microbial Resistance in Global Perspective*,
https://doi.org/10.1007/978-3-030-62662-4_8

foreign engagement to support systems change in Uganda's public health system. Through a wide variety of projects, we have tried, with varying degrees of success, to work collaboratively with an emphasis on medium-long term sustainability. The book described what we have come to identify as a key weakness in approaches to foreign engagement, especially in short-term externally funded projects, namely a tendency to identify the 'problem' as one of the knowledge deficits and attempt to solve that through fly-in-fly-out short courses delivered by foreign 'experts'.

Not only has this been seen to fail as a knowledge transfer mechanism; emphasising explicit (scientific or clinical) knowledge to the detriment of more implicit implementation knowledge is unlikely to deliver sustained behaviour change. A briefing document by Jhpiego,[1] an affiliate of John Hopkins University reports that, 'countries and donors spend significant amounts of funding on in-service training' and yet, 'traditional training approaches that focus on extended, off-site, group-based workshops have had limited effectiveness in improving and maintaining provider performance after training'. The brief cites a systematic literature review of interventions to improve healthcare provider performance in LMICs (Rowe et al. 2019) which found that 'one-time training interventions results in very low effect size'. Indeed, they report learning outcomes of this kind of training as 'low to none'. The approach specified by Jhpiego involves the use of 'low dose, high frequency' training. Rowe further contends that (any form of) training alone is not sufficient to improve quality; 'when training is combined with Quality Improvement efforts, such as coaching or supervision, the effect size is significantly greater'.

Building on our mentoring approach, we have applied this through what we have termed, 'bite-size, bedside' reciprocal learning. The emphasis on formal training in LMICs (CMEs) has also generated unintended consequences associated with encouraging absenteeism and per diemism. The endemic expectation that staff will be paid for training has a poisonous effect on everything foreign organisations and volunteers try to do. In the course of this intervention, this expectation raised its head on many occasions. The following is a typical response to the proposal to organise a CME. K4C has always provided refreshments for meetings but never pays per diems. However, when we proposed to run IPC training,

---

[1] https://hms.jhpiego.org/wp-content/uploads/2016/08/LDHF_briefer.pdf

with attention to COVID-19 preparedness, we were expected to pay staff to attend:

> People were looking forward to being paid even after telling them this is informal...

The normalisation of this expectation and the contribution such payments make to livelihoods resulted in a Ugandan colleague suggesting that we consider making financial payments in place of refreshments. The extent to which this 'poison' has begun to pollute even mentoring relationships is evident in the following conversation involving a senior doctor, who is generally absent from the ward, pressurising a Ugandan K4C health worker for money:

> He is saying that he is giving a lot of support to (highly experienced UK doctor) and therefore he needs some kind of pay for this work.

The use of per diems has commodified and distorted incentives for training. It often results in the wrong staff (often only senior or managerial staff) being trained and pollutes team relationships. Our previous book reviewed the effectiveness of these approaches in anything but the immediate term (where pre- and post-test results, as anticipated in logframe approaches, may show knowledge acquisition) when it comes to understanding behaviour change dynamics. As much behaviour change theory[2] argues, adding knowledge alone (to increase individual 'capabilities') will rarely work when the opportunities to exercise that knowledge are missing. This is particularly evident when it comes to formal training on hand hygiene, for example, when—as was the case on the PNG ward—there is no running water, soap, towels or hand gel. In such situations, training staff in standardised international protocols can be both insulting and demotivating. We also know from experience that even when the infrastructure and materials (described in behaviour change theories as the 'opportunities') to utilise knowledge are in place, this often fails to translate into hoped-for behaviour change. The reasons for this are

---

[2] Rather than rehearsing the theoretical arguments, here we refer you to Chapter 4 in our book, 'Can (imported) knowledge change systems? Understanding the dynamics of behaviour change' (pp. 79–113) available as a free download at—Mobile Professional Voluntarism and International Development: Killing Me Softly? (2017–Open Access)

highly complex. Michie et al. (2011) focus on the role that motivation plays in combining knowledge and opportunities to facilitate individual behaviour change. We felt concerned that many of these theories, rooted in psychology, tended to overemphasise the individual and gave insufficient attention to the structural contexts that crush motivation. Turning to ideas from evolutionary economics, we were captivated by the concept of 'imagined realities'. As we framed it in the book: 'Is it possible in the context within which Ugandan health workers are placed to imagine a different reality?' (p. 103). We contend that only in situations where a critical mass of people can begin to imagine sustained positive change can effective knowledge mobilisation and behaviour change take place. Harding makes a similar point in relation to research methods arguing that feminist research is, 'interested in models that stress context rather than isolated traits and behaviours. Interactive rather than linear relations, and democratic rather than authoritarian models of order' (1991: 301). The emphasis on behaviour change needs to be at the level of the organisation and organisational culture and not specifically, at the individual.

This experience framed our approach to the MSI. And to that extent, it cannot be described as strictly 'inductive' as we are continually learning from prior knowledge and associated research review. We proposed to build on our existing strong relationships with managers and local health workers established through continuous engagement and strengthened through the deployment of local and international 'volunteers'. We did not conceive of these K4C faculty (and ourselves) as the experts with solutions, but as colleagues playing a supportive role and engaging in mutual learning. We were fully aware that the expertise on the ground is of equal and often greater importance. Each of the participants, whatever their cadre or level of seniority, brings new knowledge, ideas, and enthusiasm to the table. Where specialist knowledge has been required, this has tended to take the form of multi-disciplinary encounters, of horizontal knowledge exchange with an emphasis on communication skills. Antimicrobial resistance is highly complex and, from a scientific perspective, conceptually difficult. Understanding the processes of culturing and testing bacteria for resistance and the implications of this for prescribing behaviour is not easy for those of us who are not trained in microbiology or pharmacy. Wanting to acquire such knowledge demands a real commitment and an ability to imagine a different approach to working. With this in mind, we designed an intervention based on K4C's commitment to co-presence with continuous project engagement on the ground. This creates an environment within which carefully managed and planned short stays can have a high impact and bring new interest into a project. This applies not only to visiting colleagues from the UK but also to students and interns passing through the wards. Where the need and interest are articulated and the timing is right, short stays may provide important opportunities for more formal training input. Where possible, we have conducted this in 'bite-size' chunks on the ward. And where the opportunity has arisen and the demand was patent, we have provided this through larger multi-disciplinary formal training sessions. A British volunteer pharmacist observed the success of this approach and likened it to knowledge transfer mechanisms in the NHS:

> A clear indicator of knowledge change is progress in the hospital and clear progress has been made. There is clear visibility of pharmacists on the ward – midwives know what to do with culture and sensitivity tests. The whole model is mirror like to the NHS. Clearly knowledge exchange has taken place we have just done it in an unconventional way on the ground; in a much more informal and effective way.

The following section presents interview data to illustrate the effectiveness of this approach from a variety of perspectives. We open with the views of the hospital administrator with whom K4C has had a strong relationship and who has played a key role in the partnership. Asked to outline what his experience of this short intervention has been, he immediately refers to the mentoring/co-working approach distinguishing it from other projects he has witnessed in the hospital:

> The project has made strong inroads. Your approach is very involving. The people who are on the project are working together. They are part of it. It's a real departure from many projects that we have seen where the people who come with the project go and then the project also goes. Where aliens are imposed on people; they come with their ideas and then they go. It's very involving. The midwives feel part of it. For me it has been a very exciting project and very successful – we feel a part of it. There has been a real change in people's perspectives about how they approach their practice. No little thing is ignored – the little things that we were ignoring - like you repair a tap – and yet we had the capacity to do these things, but we didn't do them until the project came.
> [Why that is the case?]

We had a laissez-faire kind of attitude when looking at institutional things – our consciousness has been awakened. Right now, they will say 'hey can we work on this?' The training that you have brought with the project – both the formal way - but you see with [K4C midwives] they have worked with the team with an in-built training mechanism that has strengthened our collaboration. They come on board and support changeovers of staff. We should be able to see change beyond the lifetime of the project. There seems to have been an integration of minds.

[What are the main benefits for you in administration?]

The project has helped our Human Resource by increasing numbers of staff on the wards and by improving staff training and guiding people on how sepsis is to be managed. The hospital now has the resource of more trained people. When they go, I still have trained staff.

The final part of the Administrator's comment alludes to the augmentation of staffing that co-working brings. It raises the question of whether the benefits have come from simply having more staff time on the ward and will be extinguished as and when that disappears. Certainly, local staff welcomed the additional support. During the project, a new in-charge was placed on the ward, in a Ministry rotation from Kampala. It is clear from the following midwife's narrative that the new in-charge had played a major role in instilling leadership by example and co-working herself:

Since those K4C people came in, they had to unite us. They just go and give good dressing. You find you are now working as a team. It is because of the teamwork those people introduced. They showed us good leadership. They would follow-up and afterwards they call the doctor and he comes. Even if they first deny we call again, we wait and again we call - even the senior doctors. The good leadership because when those people came in everyone was willing to act.

[Why wasn't there some resistance to these people?]

The way how they smell the sepsis cases. We used to say 'we are not going there' but when those people came in, at first it was SMELLING but you find them really dressing wounds. They were working WITH US hand in hand. They taught us how to do it, so we gained knowledge and they are good leaders to us.

[Where did the leadership change come from?]

They were really working alongside us hands on. We saw them dressing wounds and the in-charge she dresses wounds too, so we are all working together now. She is the in-charge, but you find her dressing wounds. This was not happening before. Before the leader said, 'I am the leader, so I am not dressing'.

The new in-charge also noticed the difference between hands-on co-working and formal training. When asked if running CMEs would have been as effective as having K4C midwives on the wards, she replies:

> I prefer 'on-the-spot-training.' This is what has changed attitudes and the fact that the K4C team are 'hands-on' working as part of the team. We have had CMEs but they heed no results. They listen, hear, but do not implement. With CMEs there is often no follow up – you need constant follow up to change behaviour. And how [K4C IPC midwife] is on the wards at times watching them hand wash and pulling them up – not criticising but making them constantly aware of the importance of hand washing both to themselves and to the patients.

Another midwife suggested that there may be some rollback if the K4C staff are removed from the ward at the end of the project:

> We would like it to stay that good. Maybe it won't be 100% but maybe 50% because when they are there, they are additional. But [they] are not just numbers. When they are there, they know exactly what to do and you know that. 'Hey, [she] is there – let me do this other part'. It is a division of labour and we work together. They know when I am on the ward then sepsis is covered!

It is interesting to note that the observation that the K4C midwives 'know exactly what to do' is not a reflection of greater initial knowledge or experience. None of the Ugandan K4C team had previously worked on AMR or sepsis management. They have, as many of the local staff have, grasped the opportunity to learn through co-working. An intern doctor expresses a view that, although local staff now have the knowledge, further support on the ward is needed to put it into practice:

> The care these ladies (K4C midwives) are doing is making a big difference.
> [But are the local staff also doing the same care or do they rely on K4C staff?]
> They would do good but not as much as there is always inadequacy of staff. You may find only 1 or 2 nurses on the ward and lots of work to do so they won't be able to dress the wounds...
> [Is that because the local staff lack time or is it skill/knowledge?]

They have the skills but there are not enough staff on the ground. One nurse can't give drugs, receive mums from theatre, manage septic wounds etc. It is not really knowledge – the big challenge is staff time but when we have nursing students around that pressure is down. Now it is holidays so that's when we have issues.

Another intern doctor observed a benefit arising from having more staff on the wards which has enabled local midwives to ensure that medication (and antibiotics) is administered on time. Timely administration is extremely important to ensure the efficacy of treatment and reduce AMR:

In terms of my observations in post-natal and gynae, I see more timely medication now; they are still not receiving it on time as desired but there has been improvement. The K4C staff have taken over the gynae and peri sepsis cases and wound dressing and it is now done on time. The workload of the local staff has reduced so that they are now able to administer medication on time.

[Was only K4C midwives doing the wound dressing now then? Was it just a question of workload?]

Now – you see swabbing and samples going to the laboratory – doing a lot of following up is very hard for us to do so now [we have K4C midwives] – they will withdraw a sample, get it tested – get results in 3 days and change the prescription and you get the patient recovering so quickly.

[So, is it having more staff or knowledge?]

I think it's a two-way thing. Let's say if the MOH gave more staff. If you had 10 more staff, I don't think they would take swabs and test them - they would go back to their old excuses. So, it's not just a reduction in workload that brings the change. It is important to note that we need people employed specifically to do that – to fight infection on the ward – dedicated people. It means our people have not appreciated the importance of doing that.

I still think if we don't have that they will go back to their old ways – sitting around pressing their phones and not doing their work. It could slide back if you removed Rachel and Dorothy. There is a solution to that. I don't know if it is poor motivation or... you see even Dorothy and Rachel are not employed as usual nurses – I think if they were not here, they would not be doing the work they are doing. It's because it is their designated job – they are employed by K4C to do this. We need to have people employed specifically for that.

At this point in the interview, he seems to suggest that the changes may lack sustainability. However, his perspective shifts as he refers to the inheritance of a 'trait' on the ward:

> Wound management has improved greatly, and mothers are receiving twice daily dressings, so they improve so quickly. The staff have developed a trait to inherit. It's good that the local staff have seen the others. They know if you have a wound and take good care of it, they will get better quicker. It's a great impact. It is even something you can push into other facilities.

In response to a question about the engagement of local midwives, he explains how one of the local midwives, *'goes the extra mile to do better. She keeps reminding me'*. He then describes the improvement in team working and the impact this has had on his own behaviour:

> Leadership on the ward – you see the medical profession is hierarchical – a system without leadership is bound to fail and I think it works now. Yes at one time Rachel did great work - I spoke rudely to her as she followed me and insisted I look at a patient who I had already passed by in the ward round – but they had the results of the tests back after I'd seen her – she kept pushing me. In the end the senior pharmacist prescribed.

He acknowledges that this pressure (from a midwife) had been a good thing in encouraging him to respond immediately to the results:

> this has been a great thing for me as a doctor – now I have seen how significant the tests are and prescribing medicines on time I think I will use that as a skill even when I leave FPRRH.

These reflections are interesting. He attributes the behaviour change on the ward not simply to having more staff but what he perceives as the allocation of specialist, dedicated, roles to staff. In practice, this was not the case. K4C did not allocate designated roles or role descriptions to our colleagues. They were placed on the ward and, with the support of a UK (diaspora) nurse who had previously undertaken specialist work on surgical site infections and wound management, they collectively undertook a scoping exercise. This identified some key problem areas which they then started to work on with local staff to resolve. Nevertheless, his proposal that staff on the ward be given specific roles may be worth considering. His contention that it is mainly K4C midwives engaging in wound management and swabbing is not borne out by the interviews, or perhaps more importantly by ethnographic observation. Asked whether all staff or just K4C staff are engaged in these activities, a midwife comments:

> We are all doing it and the laboratory is working. I believe we ALL know what to do now – we can take off the samples – arrange prescriptions – we can – we are able – we will continue – of course the main thing will now be staff shortages.

One of the pharmacists commented on the specific value of mentoring and hands-on co-working suggesting that it was not just the extra staff but the fact that they engage hands-on alongside the local staff:

> Because there are K4C personnel on the ground working with people and mentoring them. I have a strong suspicion if they were not there, we would not have seen this change – showing people how to do things and mentoring them. They are hands-on and people see what they are doing – how they are cleaning wounds and people will follow their lead – they see them actually doing the work and feel motivated to join in. You see them criss-crossing up to the laboratory. If you don't have people on the ground, you won't get accurate information – I've noticed that, so presence on the ground is key.
> [Would the same change have come if we had UK volunteers]
> You need people on the ground for a long period. The perception if there are British people, they see whites as only coming for a short time. They are there but anyway they will go away so they would rather have a local person who stays longer.
> [What about [K4C nurse volunteer]? She came for a short time?]
> Well they saw her as local – she was accepted as a local. Even me I thought she was Ugandan.

As we noted above, the mentoring relationships on the ground are complex and fluid with knowledge exchange taking place in a variety of directions. The greater engagement of the pharmacists on the ward has created new opportunities for active mentoring within the pharmacy team itself. One of the senior pharmacists describes this process:

> The interns are happy to do the clinical work now. It is challenging, motivating, and inspiring but they appreciate my presence and my authority on the wards. It has opened their eyes to the possibilities in their professions. Most pharmacists tend to be frustrated when they qualify and there

is nothing for them to do in the hospital apart from signing requisition orders... our system does not support our role. It gives them confidence.

The decision, by the project team, to employ one of the pharmacy interns when his internship came to an end mid-project has stimulated a number of mentoring opportunities. In the first instance, it has freed up the hospital pharmacist to engage directly with staff, students and interns on the wards. Second, it has extended a period of more intense mentorship and enabled him to get more involved in leadership and research:

> [He] was on his way out but he has benefitted from that – he has got what his colleagues didn't get. Clinical pharmacy has to be practical; you can't sit in the classroom. You have to have the practical relevance and you need someone who inspires and encourages you.

The expectation that projects such as this must involve formal, off-ward CMEs, was expressed by a senior pharmacist visiting the hospital on behalf of the Commonwealth Pharmacy Association who insisted on knowing what formal training the pharmacy interns had received. The senior hospital pharmacist replied:

> We have had interactive, informal, training. The people on the ward now ask for the pharmacists – there is real behaviour change – they didn't used to ask.

It is interesting to consider how the K4C staff reflected on their own roles on the ward and the impact of these. One of them used the word 'training' when talking about local staff. This triggered a question about what that involved:

> [You refer to them doing 'what we trained them to do'. When did we train them as such?]
> Well when I do a procedure, I work with one of the staff. We work together to do the procedure. Wound dressing is a two-person job, so we work side by side.
> [How do you do that? If a wound needs dressing do you go and ask someone to help?]
> Yes, I just go and ask someone to come and help me. Sometimes we assist them. It doesn't have to be us doing it all the time. Most of the time you need an assistant, so you don't contaminate things around you

– so each time one person is doing the dressing and the other person is an assistant.

On rare occasions when there is only one staff on the ward you find you're doing it yourself but like yesterday we found a nurse doing a wound, so I went along to help. She doesn't see it as a threat, we work together. It's the way people come on board that matters because if you come and they see you as one of them … but if you come at another level and they look at you as telling them what to do – the working relationship would be poor – me I see them looking at us to help. As long as were there to give a hand they will usually come on board.

As noted previously, the K4C team were not themselves experts when the project commenced. They describe how they have learnt through a process of mentoring within a multi-disciplinary team:

It is like a continuous update. When I was in (college) training, they did not go into details about taking samples. Now we have more knowledge about it, we know how we are doing it and we can interpret the results and see how to get these isolates causing this problem so somehow, we are getting knowledge.

[Did you have that knowledge at the start of the project?]

Not as such - somehow - but I know a lot more now about the microbiology

[So, how have you learnt then?]

Because we have multi-disciplinarity. If it is a new isolate, we can ask in the laboratory and the pharmacy. If the antibiotic is not working, we can look at alternatives. Then you get to know.

[So, you have learnt as a team but with them]

Yes, we get knowledge from different cadres – from the doctors – the laboratory – we have learnt together. And [lab scientists] trained us as a team on how to take cultures and how to interpret them – he trained us on the ward. They showed us how to take sterile swabs and avoid contamination - we have also not got any contamination in our swabs – according to the results from the laboratory.

Also, because they are changing antibiotics, that is a group thing, so we learn as a team. We discuss it. We never find just a pharmacist changing the drug – we all discuss it together. Like when we discussed a patient reacting to doxycycline, we all discussed it together. So, it is a multidisciplinary team.

[In the past would you have started to ask such questions about antibiotics?]

No, but as soon as it started it was a multi -disciplinary approach.

When you get this patient on the ward, we pull everyone together and discussion takes place. People realise it is not a one-man job. It is more interesting as most of the time the interns are on the ward, so everyone realises this is something we have to do. Ourselves, this has been new to us. Now there is a lot coming through – now we see the results of teamwork. Everyone is concerned about a patient wasting away and this has motivated everyone.

Knowledge Transfer and Behaviour Change

Harding's work on objectivity in research describes how development policies targeted at the Global South are imbued with concepts of knowledge deficit and, as a result, failed to achieve their objectives. The 'strong objectivity' approach, on the other hand, she suggests has been more effective:

> The transfer of Western scientific rationality and technical expertise from the West to "the rest" had always been the "motor" of modernisation theory and now drives development policy [...] many of the assumptions about women and poor people in the Global South - were false..[..] the strong objectivity approach enabled the goals of improving to the living conditions for poor women and of decreasing poverty overall to be advanced... (2015: 152)

We have noted in previous work (Ackers and Ackers-Johnson 2016) the tendency of international development funders to assume an immediate relationship between behaviour change and knowledge. Describing this as an example of 'fetishizing training', we illustrated the complexity of apparently simple interventions (in that case operationalising triage) and the associated behavioural change required. Antimicrobial resistance is a far more complex concept than triage. Little was known about AMR until very recently and the language used prohibits multi-professional engagement. Although the science is complex and still emerging, the necessity of achieving effective science communication is paramount. It has been inspirational to hear local midwives articulate questions about AMR in sophisticated scientific terms:

> Then of course with the culture and sensitivity – right now we are doing OK. You have guided us on that one.
> [Did you know how to do it before?]

Not really. We knew how to take swabs, but we didn't know how to interpret the results. I have really begun appreciating; now we know which organisms we are fighting and can give the required antibiotics.

[So, have you had to learn a lot about these organisms and the antibiotics?]

Yes, we have had to learn a lot. The major thing I've learnt is about how mothers here are resistant to most antibiotics. Before I would hear but I had not seen it myself but at least now I have observed. Now there is a case who is only sensitive to amikacin.

THE CONTRIBUTION OF FORMAL TRAINING IN A MENTORING CONTEXT

In keeping with an inductive approach, the MSI did not commence with formal training or efforts to introduce externally produced tools or protocols but with on the ground relationship-building and co-working. Towards the end of the project, we have organised more formal training usually in a small 'doctors' room on the ward or in the larger hospital boardroom. Semi-formal short CMEs on topics such as hand hygiene or management of the newly organised evacuation room (and associated cleaning and sterilising procedures) have been bolted onto to clinical audit meetings which, since the project, occur every time there is a maternal death or a complex case for discussion. During a project visit in January 2020, the team responded immediately to requests from local midwives and intern doctors to improve their ability to interpret laboratory results and provided a short ward-based CME. Strictly speaking, midwives don't 'need' this knowledge; their role in theory is to handover the results to the doctors and for pharmacists and doctors to deliberate and prescribe. The desire to know more and understand the processes which had led to such remarkable improvements in patient outcomes is an indicator of the motivational impact of change and how that fosters a quest for knowledge.

Somekh (with reference to educational learning in the UK) does not use the language of behaviour change that has become so popular in recent years but links learning closely to notions of personal efficacy and autonomy: 'learning is closely related to a sense of personal efficacy – to learn children need to have autonomy and take responsibility' (2006: 4).

This has echoes of the idea of imagined realities and the belief that things can change, and you can be a part of that change. Defining 'agency' as the 'capability of a self to take actions that will have an impact on a social situation', Somekh contends that change processes need to, '*unlock the agency of individuals and groups so that groups work interactively and reflectively to go beyond their personal learning and aim for a broader impact on improving working methods and practice across the whole workplace*' (2006: 21).

Of course, a sense of personal agency is inextricably contingent on the context (although we would argue that there is always at least some potential for agency). Whilst behavioural scientists may use the more benign concept of 'opportunities' to capture context, social scientists have articulated this relationship in terms of structuration theory or structure-agency models that account in more complex ways for the webs of constraints that shape individual behaviour.

The final chapter sums up the achievements of the Maternal Sepsis Intervention approach with a discussion on how you capture change processes in complex interventions; how you sustain those changes and how lessons can be applied to other settings.

References

Ackers, H. L. & Ackers-Johnson, J. (2016). *Mobile professional voluntarism and international development: Killing me softly?* Palgrave PIVOT http://link.springer.com/book/10.1057%2F978-1-137-55833-6.

Harding, S. (1991). *Whose science? Whose knowledge? Thinking from women's lives.* Ithaca, NY: Cornell University.

Harding, S. (2015). *Objectivity and diversity: Another logic of scientific research.* London: University of Chicago Press.

Michie, S., van Stralen, M. M., & West, R. (2011). The behaviour change wheel: A new method for characterising and designing behaviour change interventions. *Implementation Science, 6*(1), 42. https://doi.org/10.1186/1748-5908-6-42.

Rajkotia, Y. (2018). Beware of the success cartel: A plea for rational progress in global health. *BMJ Global Health, 3.*

Rowe, S. Y., Peters, D. H., Holloway, K. A., Chalker, J., Ross-Degnan, D., & Rowe, A. K. (2019). A systematic review of the effectiveness of strategies to improve health care provider performance in low- and middle-income countries: Methods and descriptive results. *PLoS ONE, 14*(5), e0217617.

Somekh, B. (2006). *Action research: A methodology for change and development.* Maidenhead: Open University Press.

CHAPTER 9

Conclusions: Capturing, Sustaining, Spreading and Researching Change

Abstract This chapter summarises the overall impacts of the Maternal Sepsis Intervention reflecting on the processes of capturing, sustaining and spreading best practice in antimicrobial stewardship. It argues that sustainability is achievable when an intervention reduces hospital costs and, in such cases, the responsibility or sustainability rests with LMIC institutions. Key barriers to sustainability include supply chain weaknesses and human resource limitations. The chapter recommends the use of Public–Private Partnerships to help to overcome these barriers.

Keywords Sustainability · Public–private partnership · Policy transfer · Snagging

Dyar et al. (2017) review the use of the term 'antimicrobial stewardship' and conclude that there has been an overemphasis on conceptualisations focused on 'individual prescriptions' and insufficient emphasis on the societal implications of antimicrobial use. Furthermore, and of particular relevance to the MSI project, there has been insufficient translation of the concepts of 'responsible use' into context and time-specific actions. The authors conclude that AMS is not so much a concept as it is a tool to assess whether organisations are identifying actions to improve responsible use in the specific context within which they are functioning. This idea fits

L. Ackers et al., *Anti-Microbial Resistance in Global Perspective*,
https://doi.org/10.1007/978-3-030-62662-4_9

very well with the action-research approach used in our intervention and the results arising from that.

The final chapter includes three sections. The first section 'Capturing Change' summarises the effectiveness of the intervention identifying key outcomes and the drivers that have contributed to these. The second section 'Sustaining Change' addresses the issue of sustainability of those change processes, and the third section 'Spreading Change' discusses the opportunity to translate and transfer aspects of the approach within the hospital, throughout the country and to other international contexts. We draw the chapter to a close reflecting on the benefits of the ethnographic approach and the attention to contextual detail and co-researching that this has afforded.

CAPTURING CHANGE

Although we were acutely aware of the importance of Infection Prevention and Control (IPC), the emphasis in the CwPAMS call on more complex concepts of AMR and AMS distracted us somewhat from the foundational importance of IPC. Poor attention to IPC in the hospital clearly contributed to high levels of protracted and expensive hospital-acquired infection creating more and more unpleasant work for health workers. IPC improvement has proved to be a key driver of behaviour change. Health workers have viewed the changes as an immediate investment in their own health and safety, improving their ability to work effectively in a safe, supported environment. As such IPC has not just improved the 'opportunity' component of behaviour change, it has engaged directly in health worker motivation. Attempting to address AMS without full attention to IPC would, we believe, have failed to capture the imagination and commitment of ward staff. Capability (or lack of knowledge or skill) did not emerge as a key factor in behaviour change, and where knowledge mobilisation has taken place, it has occurred primarily through mentoring and role modelling with some attention to specifics (like when hand gel should be used in relation to washing or what concentration of JIK should be used to clean mattresses). Nurses and midwives have emerged as the natural leaders in these areas playing a key role in shaping organisational culture and orienting rotating doctors and students. The financial costs of implementing major improvements in the IPC infrastructure have been minimal in comparison with the savings achieved through shorter hospital stays, reductions in re-admissions and use of theatre for secondary wound closures.

Change has come about over time through a myriad of small changes and attention to the limitations or unintended consequences of these. The process has generated active and positive dialogue and collegiality. Once the 'obvious' issues around hand hygiene and equipment sterility had been resolved, the management of attendants emerged as a key and stubborn weakness in IPC systems. Whilst progress is being made and has been accelerated in the COVID-19 environment, further work needs to be done in this area throughout the hospital and at national level. The impact of stock-outs on the ability to maintain basic hygiene on the wards is another structural issue requiring intervention at the level of national and institutional supply chains. We return to this issue below.

The IPC work focused attention on wounds and wound care on the wards. On one level, this was very much an IPC issue; improving IPC reduces the risk of wounds becoming infected on the wards in the first place. Managing those that do become infected (or arrive infected) reduces the risk of sepsis and the need for secondary closure. Cleaning and dressing wounds emerged as a key, if unforeseen, dimension of infection control; it also provides the opportunity for swabbing and, with that, for complex multi-disciplinary 'huddling'. At this point, the nurses and midwives hand the baton to the laboratory and microbiology results create a unique opportunity for multi-disciplinary dialogue grounded in scientific results. Once these results are available, so the opportunity for clinical pharmacy emerges with pharmacists engaging actively with intern doctors to support evidence-based, 'rational' prescribing. The overwhelmingly positive impacts of these processes on patient outcomes, the environment on the ward and the satisfaction at being able to save lives and shorten hospital stays feeds back into the loop creating an active momentum for change. This in turn generated an appetite for knowledge, to breach disciplinary boundaries, share a new language of communication and experience mutual respect.

The findings evidence significant and impactful behaviour change on the wards with genuine multi-disciplinary team working contributing to changes in prescribing behaviour and AMS. Wound care and laboratory testing lie at the heart of these changes centre staging nurses, midwives and laboratory scientists in stewardship processes. The results of the microbiology testing then provide a platform for genuine multidisciplinarity and, specifically, the first opportunity for clinical pharmacy engagement. This is true both at the level of rational (evidence-based) prescribing for individual patients and in improving the evidence base behind empirical prescribing (through an understanding of resistance patterns). If access to a full range of antibiotics were available, this platform of behaviour change would transform antimicrobial use patterns, reduce the over- and inappropriate use of antimicrobials, improve patient outcomes and deliver significant cost-benefits.

The changes described above are essential to achieving 'responsible use' in a Ugandan public hospital; but they are not sufficient. Cox et al. make the important point that, *'delayed or no access to antibiotics kills more people than antibiotic resistant bacteria. … AMS is not only about reducing inappropriate use, but also assuring access to effective treatment'* (2017: 813). The changes create the opportunity for active pharmacy engagement in multi-disciplinary decision-making. This achievement could have a major impact on antibiotic use as post-natal and gynae wards consume by far the largest volume of antibiotics at the hospital. The cost-effectiveness of this intervention underlines the sustainability potential and the immediate opportunity for scale-up across the hospital as a whole and to other public health facilities in Uganda and beyond.

Chapter 6 has elaborated the complexity and opacity of the supply chain system in a public hospital setting in Uganda. In the absence of effective supplies not only will these cost savings elude facilities; the patients involved will fail to thrive and the motivation of health workers to apply the skills and knowledge they have demonstrated will inevitably decline. We have explained in detail the complex dynamics of supply chain management. Understanding and piecing together these processes have required painstaking ethnographic research to unpick major errors in record-keeping and interpret the trends observed. In the first instance, the very centralised system creates huge dependency on the functionality of National Medical Stores and the adequacy of Ministry budgets. Centralisation may be seen as necessary in systems so damaged by corruption but where this undermines flexibility and responsiveness and generates extended and predictable stock-outs, the systems put in place to improve antimicrobial stewardship will, inevitably, fail.

The lack of supervision of rotating intern doctors by their senior colleagues restricts the effectiveness of the approach and the osmotic potential of continuous multi-disciplinary engagement. Nurse and pharmacy leaders have filled the knowledge gap through orientation and mentoring of intern doctors demonstrating the potential of task shifting in practice. Further improvement to prevent surgical site infection requires attention in the operating theatre as women undergo caesarean sections or laparotomies and in complex secondary wound closures. The deployment of a UK doctor in theatre had a very positive effect often substituting for the absence of senior doctors. Ultimately, this is unsustainable, and measures need to be taken to ensure the constant presence of senior doctors in obstetric theatre and in a supervisory capacity in public hospitals.

The change mechanism deployed in the MSI is grounded in a particular approach to knowledge mobilisation and behaviour change. Building on extensive experience of project operationalisation, the team was acutely aware of the limited efficacy of fly-in-fly-out formal training programmes organised by foreign teams and lubricated by opportunities to leave the ward or receive per diems. We have established the working principle of co-presence with colleagues working side by side on the wards in more democratic and genuinely bilateral[1] mentoring relationships. Where Ugandan health workers have articulated a desire to gain new knowledge, the project has mobilised to provide that. In some cases, this has involved short sessions led by UK experts—such as the interpretation of laboratory results, for example. In others, the whole group of pharmacy interns presented a workshop on AMR to the hospital. Horizontal knowledge mobilisation between different disciplinary groups is a feature of this process rather than a comparison of protocols or practices in high- and low-income countries.

Co-presence is not only a principle to be applied to health worker training; it also forms the basis of ethical and equitable approaches to co-researching.

Sustaining Change

Sustainability has become a key concept in overseas development interventions. As with most things, sustainability needs to be understood in context; it is also highly contested. Sustainable Development is defined

---

[1] The team have been allocated Commonwealth Fellowships to support 5 Ugandan health workers to come to the UK for 3 months to develop their learning around tissue viability and wound management.

in the 1987 Brundtland Report (United Nations 1987) as: 'Development that meets the needs of the present without compromising the ability of future generations to meet their own needs'. Thwink.org[2] provides a very simple definition that resonates with the objectives of the MSI, describing sustainability as, 'the ability to continue a defined behaviour indefinitely'. The powerful emphasis on sustainability in international development has arisen largely due to increasing recognition of the ineffectiveness of aid and, in some cases, the damaging externality effects associated with aid dependency (Moyo 2010). We have noted the limited evidence of the impact of many foreign-supported training initiatives including the many that have failed to mobilise knowledge from theory into practice and impact behaviour. The MSI has demonstrated genuine and sustained behaviour change in Infection Prevention Control and rational prescribing. But can this behaviour be sustained?

Implicit in many discussions of sustainability is the notion of an 'exit-strategy' to guard against funding dependency. Whilst we are very aware of the damaging effects of foreign aid dependency, we would argue that exit is not a zero-sum game. Continuity of relationships and partnership has been the hallmark of K4C's engagement in Uganda. It is the quality of transitions from a phase of more intense engagement to a more routine maintenance that is of importance here. Many aid interventions are characteristically brief resulting in rapidly designed interventions followed by immediate withdrawal. The CwPAMS funding is typical in that regard; the funding period was extremely short with no lead time and discrete measurable and attributable outcomes expected within 15 months. The co-creation and co-working model has ensured high levels of local engagement and genuine team working both within the hospital and between project partners (in Uganda and the UK).

SUSTAINABILITY AND HUMAN RESOURCE AUGMENTATION

There can be no doubt that individual capability (knowledge and skill) to improve AMS exists amongst all cadres. In many areas, this existed prior to intervention particularly in relation to a fundamental understanding of infection prevention control and wound management. Health workers also referred to gaining new, more specialist, knowledge and

---

[2] https://www.thwink.org/sustain/glossary/Sustainability.htm.

implementation skills which would ensure continuity and were clear that the work would continue after project funding ends. The feedback from ward midwives evidences their recognition of the new knowledge that has come with the project and their belief that they are now able to maintain the behaviour change on the ward. However, they all expressed concerns about the impact that staff shortages will have on their ability to deliver a quality service. In some respects, the project has increased the complexity of tasks on the ward; midwives are now more actively engaged in new tasks such as swabbing; managing the process of sending samples to the laboratory and receiving them (which used to be the role of the doctors); and then pulling the teams together to act on lab results. There is also an increased burden on the staff and the in-charge to manage patients and visitors and ensure all new students and interns are effectively oriented in the culture and practices of the ward. There is an important task-shifting component here as the in-charge and pharmacist are effectively fulfilling the role of the senior doctor in orienting intern doctors. On the other hand, measures to enable midwives to work more effectively, reducing frequent visits to surgical theatre for sterilising equipment and gauze and physically visiting the lab for test results, have reduced demands on staff. In reality, the K4C staffing team settled into a routine of having one person on the ward every day of the week (mornings only at weekends and no nights). The costs of this additional and perhaps dedicated midwife could be considered money well spent by the hospital if they are keen to maintain the model and the cost savings it has brought.

The success of the MSI is not contingent on the skills or knowledge of short-stay foreign volunteers, or on foreign income sources. Critically, the intervention has been recognised by the hospital as cost saving:

> When sepsis is managed then the resources used to manage these patients reduce drastically. By reducing long stays which brings about savings. You have contributed to the hospital budget with real term savings (Hospital Administrator).

The findings evidence the overwhelming success of the approach in terms of reducing infection risks; managing infections more effectively; improving prescribing behaviour; and reducing maternal morbidity and mortality. In the process, the evidence of reductions in patient stays, use of operating theatres and readmissions represent significant cost savings for the hospital. Asked about the sustainability of the model at an IPC meeting, Professor Ackers replied:

> Everything we have done here is completely sustainable. We have not put in big fancy investments or one-off trainings. We have saved the hospital considerable funds; the responsibility for sustainability now rests with the hospital. I don't think having one extra member of staff is too difficult for the hospital to maintain. We have not brought in some big thing that we are now taking away.

Indeed, to push this point further, an extension of the MSI approach to other wards in the hospital would result in further cost savings. In this context, where an intervention has improved outcomes whilst also reducing costs, the responsibility for sustainability rests with the hospital and health system to put systems in place. This would imply some investments and delegation of national autonomy—particularly in employing more nurses/midwives in place of the K4C staff and more pharmacists to support clinical pharmacy on other wards. These measures are within the hospital plans and jurisdiction and fully actionable. More detailed exposure of costs may lubricate the politics of internal decision-making. A discussion about the mechanics of undertaking meaningful and accurate cost-benefit analysis to identify the financial impact of the project has taken place with hospital management. Certainly, accurate assessment of costs would facilitate evidence-based human, and overall, resource management. And failure to make these decisions and augment staffing to protect the model, post-project, will undoubtedly result in cost escalation as patient stays begin to increase. The Health Partnership team are committed to supporting this process. There can be no doubt that managing the volume of patients on the wards presents serious challenges. One option for the hospital management to consider is the use of creative financial management to reinvest a portion of the cost savings achieved through the intervention model to employ additional staff.

The impact of damaged referral systems and the ability of referring hospitals and health centres to bump patients on to referral hospitals with impunity and take no responsibility for the mortalities and morbidities they contribute to require the engagement of the Ministry of Health.

SUSTAINABILITY AND ACCESS TO SUPPLIES

Access to supplies, in relation to IPC, wound management and antimicrobials, has proved an ongoing concern. Although substantial improvements have been made, these are sub-optimal and realising the full impact of the approach depends on improving access to supplies. Aid dependency is not simply about the size of financial commitments in relation to local income; it is also about the mode of engagement. The concept of 'conditionalities' has been deployed in an attempt to reduce no-strings-attached 'giving' or 'donations' which undermine sustainability. In general, simply 'donating' consumables can never be sustainable and it can and does encourage new forms of corruption. It is for this reason that K4C as a charity does not routinely engage in these forms of support.

We are acutely aware that our ability to make even very small payments can save lives. During the MSI, we made one such payment (of £30) to purchase amikacin when we knew that, in the absence of that payment, the mother would almost certainly die. The problem here is not just about costs; it is structural. Amikacin is not currently available through National Medical Stores, and as the hospital budget for drugs is held at NMS, the solution requires more structural changes which we hope to support through active advocacy work (see below). K4C also gave the ward two bottles of JIK when supplies ran out. This cost £6 but safeguarded the improvements in basic IPC on the ward. The problem here was caused by NMS stock-outs. K4C has also been supplying hand gel to the ward for the duration of the intervention. The decision to engage in what looks like unsustainable investments was made cautiously; K4C has the ability to manufacture hand gel locally and the costs of doing so are quite low.[3] The decision was made to provide it given the total lack of supplies and the fundamental importance of hand hygiene to the achievement of behaviour change.

Concerned about the ability to sustain hand gel production in the previous project (funded by the Tropical Health and Education Trust), K4C proposed a Public–Private Partnership agreement with FPRRH. K4C generates income to sustain its own activities between project funding, through its student placement programme (Ahmed et al. 2017). In most hosting facilities, K4C makes investments, based on Fair Trade Principles, including the provision of staff and infrastructure, etc., on

---

[3] The costs of providing hand gel to the PNG during the project came to around £100.

a pro-rata basis in recognition of support for the placements. FPRRH had insisted on cash payments (of £150 per student placement) rather than the in-kind investments deployed in other health facilities. K4C had proposed that at least some of this payment could be made into a PPP account, especially as K4C provides supervision for students through its own staff co-working on the wards. This income could then be utilised to underwrite the costs of IPC. At the time, this proposal was rejected by the hospital director. The proposal was raised again during the MSI and the start of the COVID-19 pandemic. At this point, K4C was keen to support the hospital to put IPC systems in place across all its wards. A Public–Private Partnership agreement has now been signed generating a more sustainable and integrated mechanism for supply chain augmentation with an emphasis on IPC and antimicrobials. This provides a unique opportunity for foreign organisations to cooperate on a co-decision and co-funding basis, guided by the hospital's Medicines Therapeutic Committee and supported by a reputable not-for-profit supplier, Joint Medical Stores. The objective will be to move away from dependency-generating donations to a more integrated approach with the agility to respond to local needs.

O'Neills review of AMR (2016) identified a number of 'guiding principles' for implementation. These included an emphasis on cost-effectiveness and the identification of 'market failures affecting resource allocation, regulation and price mechanisms'. Most of the solutions advocated in his review engage at international level; the proposed PPP mechanism we identify operates at micro-institutional level[4] and we hope will become a scalable model. This represents an important opportunity to improve sustainability at the local level. Detailed evaluation of this mechanism will contribute to a wider dissemination strategy that will support advocacy work at national and international levels.

[4] The MOH, Uganda has recommended to development of PPP as part of its Health Strategic Health Policy. This is echoed in the National Medicines Policy (2015: 30) that promotes PPP in the pharmaceutical sector.

Spreading Change

The MSI has demonstrated the potential for change and the efficiencies associated with this. We hope that publication of this evidence will stimulate discussion at national level amongst all key stakeholders and generate a momentum for change.

Policy Transfer Opportunities Within the Hospital

The most immediate opportunity to scale up and apply the MSI approach lies with the hospital itself. The Infection Prevention Control (IPC) Committee has already played an important role in exposing key players in the hospital to the changes in practice introduced on the post-natal and gynae wards. It is interesting to note that some of the concerns expressed on PNG wards about the fluidity of intern rotations have created an interest in and appetite for change as health workers who have spent time on PNG subsequently move to other wards. From a governance perspective, further work needs to be done to fully establish and institutionalise the hospital Medicines Therapeutic Committee (MTC). This process has been complex and circuitous. The pharmacy team acknowledged the central importance of the MTC in policy implementation processes. Whilst FPRRH is one of only very few RRHs in Uganda to have functionalised the Rx medicines management system, the pharmacy team expressed concern that, in the absence of an MTC, the full impact of this could not be realised:

> Before we didn't have the people to operate the Rx system. It helps us know about the supply chain and which drugs are about to expire. In theory we should be doing reports and submitting them to the MTC, but the committee is still not functioning.

There is also concern that some of the measures developed are not fully recognised by hospital doctors because they have not been validated through an established MTC:

> The 'Antibiotic Selection Tool' we had designed has been removed from the wall; we think it is because it has not gone through the MTC.

This illustrates the importance of the MTC to the validation and endorsement of policy and medicines management. Whilst considerable efforts have been made on the post-natal and gynae ward to block the prescription of high-end antibiotics without lab tests, extending this policy across the whole hospital requires MTC endorsement. The project co-lead and Secretary General of the Pharmaceutical Society of Uganda expressed a similar concern about the role of the MTC as a platform for the utilisation and dissemination of data and the institutionalisation of the model used on post-natal and gynae wards. We propose a stepwise cautious incremental approach to the scaling up in other wards commencing in either surgical ward or paediatrics. The COVID-19 pandemic has thrown a light on supply chain efficacy and the impact of weak supply chains on global and national inequalities. Although the poorest in societies will suffer disproportionately, the tentacles of antimicrobial resistance, as with all global pandemics, will reverberate across the globe.

There is a growing concern embodied in evolving research governance that ethical research demands attention to dissemination. This includes active engagement with stakeholders and policymakers and aligns with the principles of partnership commitment. This also extends to a commitment to accessibility as Harding suggests:

> Research should not be elite [..] it would have to be accessible – physically and intellectually to anyone interested. It would be humble and acknowledge that each new "truth' is partial; that is incomplete as well as culture-bound' (1991: 300).

Open access publishing plays a key role in this process; it also demands attention to the use of language to ensure that all actors in multi-disciplinary audiences are able to make sense of research whilst also not over-simplifying key messages.

Researching Change: Ethnography and 'Snagging'

Providing the evidence base to inform and facilitate behaviour or systems change processes is notoriously complex even when researchers are working in their own country and institutions. Add to that the complexity of power relations and the dynamics of organisational culture in very different national contexts, and it is easy to understand how interventions in global health can fail. Co-working and co-researching have combined in this study through a form of action-oriented ethnography. Ethnography, as a profoundly inductive and intuitive approach has enabled a whole series of small 'snags'[5] or 'knots' to be identified and responded to with a level of agility that has enabled the intervention to move forward in a reflexive process. And this has made people feel that the intervention and research have been attentive; that their views have been listened to, respected, and responded to. Challenging AMR demands attention to detail, to the minutiae and mundane features of health workers' everyday lives.

The next section reflects on some of the prescriptions and assumptions that shape the global and national response to AMR and are echoed in the CwPAMS 'positioning' that framed the funding programme. We outlined the Ugandan National Action Plan—itself an implementation device of the Global Action Plan. Certainly, all five areas resonate with our intervention findings as key priorities. Whilst it is valuable to distinguish the areas, they are essentially indivisible and interrelated. With the exception of surveillance, where scientific methodologies can help shed light on the mechanisms, transmission patterns and origins of bacteria and resistance, we would argue that any intervention must span all 5 areas. Cox et al.'s review (2017) identifies the lack of evidence on AMS derived from highly contextualised interventions. The focus on maternal sepsis in our work will, we hope, represent a response to that call with context here not referring simply to work taking place in LMICs but to work closely engaged with health system priorities and a patent local need. Embedding AMR/AMS work in this way creates a powerful resonance and tangibility about the impact of what are, in many ways, quite intangible concepts. The CwPAMS call, with funding of £1.3 m, represents a 'drop in the ocean' in the wider AMR funding landscape both in terms of UK funding to LMICS and that of other foreign actors. On that basis, we can understand the desire for focus. There is much discussion about partnerships in intervention delivery but less attention to partnerships between funding bodies. In many respects, the growing tendency for funding bodies to come together to support 'real-world' problems and combine interventions with greater attention to evaluation is to be welcomed. Disciplinary and professional boundaries have generated myopic impressions of the world. However, there are some unintended consequences of these partnerships that demand careful and honest reflection. Funding is fundamentally a political process. Chapter 2

[5] The term 'snag' is often used in building contracts to refer to 'a problem or difficulty that stops or slows the progress of something'.

referred to the dominance of certain paradigms in global health and the heavily prescribed nature of many funding calls. Inevitably, these lead to a deductive approach shaped by powerful normative assumptions. Shiffman discusses the role that power plays in shaping global health agendas. At one level, power shapes health priority-setting and this is clearly seen in instruments such as the Sustainable Development Goals. Reflecting on the role that the Lancet has played in global health agenda-setting, Shiffman poses the question: 'Why do some individuals become recognised as global health experts'? (2014: 297). We have noted the dominance of medical and pharmaceuticalisation paradigms that have come to shape discourses around AMR and global health more generally. Our work has challenged these paradigms both from a methodological perspective (what constitutes knowledge in this context) and from a task-shifting (clinical) perspective (which cadres have the greatest potential to support antimicrobial stewardship in LMICs?).

The team have worked with THET for many years and we share many of the values embedded in the principles of health partnership working. We feel these align closely with an action-research and fundamentally inductive approach and support the concept of locally grounded, highly contextualised, case studies. They also support the need for more demo-cratic knowledge mobilisation mechanisms formed around continuous mentoring and co-working relationships. This concurs with the ethos of the Sustainable Development Goal 17 (global partnerships). The Depart-ment of Health and Social Care is a very different organisation with its primary objectives, necessarily (and rightly in our view) focused on the needs of the UK in what are very challenging times. AMR is posing a genuine and immediate threat to the UK economy and the health and well-being of its citizens. It goes without saying that any funding committed to LMICs must at least demonstrate 'mutual interest'. Our own previous work funded by Health Education England (Ackers et al 2017) reported on the benefits to the NHS from professional volun-teering identifying gains in key areas such as leadership, communication and managing in resource-constrained environments. The Common-wealth Pharmacist Association, as a body representing the interests of pharmacy as a profession inevitably and necessarily, has vested interests too. The CPA describes its role in terms of: 'Empowering pharmacists to improve health and wellbeing throughout the Commonwealth'.[6] Once again, this is an important and much needed perspective. There is substantial under-investment in pharmacy in Uganda, so much so that it is impossible for pharmacists to utilise their clinical pharmacy knowledge and skills in public facilities. Our ongoing engagement with FPRRH made us very aware that this was the case prior to project inception and the impacts of this emerged over the project, as did the potential for change when employing just one more pharmacist to support the work. What became profoundly clear though was that engagement of clinical pharmacy in the ways that the CPA planned required the engagement of laboratory scientists. Without laboratory, testing the potential for clinical pharmacy on the wards is severely compromised. And the opportunity to take samples for laboratory testing rested on the respect for and active co-working with those cadres that are continually present on the wards. In the Ugandan context, this means midwives and nurses.

The funding bodies necessarily had their own, political, objectives arising from the need to demonstrate impact on an increasingly cynical and cash-limited environment. These objectives combined in complex and 'clunky' ways to make project specification, management and evidence-generation extremely difficult. The final component of the AMR NAP (and GAP) refers to research. Good research needs space to define and change priorities, and research excellence is increasingly associated with 'impact'. Evaluation is research and should be distinguished from accountability mechanisms. It is interesting to note that little, if any, of the log-frame 'evidence' we were required to provide at regular intervals to the funding bodies has been used in this book, and little, if any, of the key evidence presented in the book could be squeezed into the log-frames.

The experience of managing this project has been characterised by ongoing creative tension with the funding bodies and managing agents bound by (and required to police) highly prescriptive, inflexible and deductive log-frame systems. Somekh describes 'episodes of substantial friction' as 'the starting point for deeper collaboration [...] co-labouring with partners involves encountering many 'knots' associated with discomfort, difficulty and frustration' (2006: 23). Friction, she argues, reflects the deep seriousness that partners attach to their work. We hope that

[6] https://commonwealthpharmacy.org.

the outcomes achieved will encourage a debate in Overseas Development Assistance work to allow projects and researchers to breathe, innovate, show the humility to acknowledge intervention failure and prepare to be surprised.

REFERENCES

Ackers, H. L., Ackers-Johnson, J., Tyler, N., & Chatwin, J., (2017). *Healthcare, frugal innovation, and professional voluntarism: A cost-benefit analysis.* Palgrave PIVOT.

Ahmed, A., Ackers-Johnson, J., & Ackers, H. L. (2017). *The ethics of healthcare education placements in low-income countries: First do no harm?* Palgrave PIVOT.

Cox, J. A., Vlieghe, E., Mendelson, M., Wertheim, H., Ndegwa, L., Villegas, M. V., Gould, I., & Levy Hara, G. (2017, November 1). Antibiotic stewardship in low-and middle-income countries: The same but different? *Clinical Microbiology and Infection.* Elsevier B. V.: 812–818. https://doi.org/10.1016/j.cmi.2017.07.010.

Dyar, O. J., Huttner, B., Schouten, J., Pulcini, C., ESGAP (ESCMID Study Group for Antimicrobial stewardshiP). What is antimicrobial stewardship? (2017) *Clin Microbiol Infect, 23*(11): 793–798. https://doi.org/10.1016/j.cmi.2017.08.026.

Harding, S. (1991). *Whose science? Whose knowledge? Thinking from women's lives.* Ithaca, NY: Cornell University.

Ministry of Health, Uganda. (2015). *Uganda national medicines policy.*

Moyo, D. (2010). *Dead aid: Why AID is not working and how there is another way for Africa.* London: Penguin.

O'Neill, J. (2016). Tackling drug-resistant infections globally: Final report and recommendations. *The Review on Antimicrobial Resistance,* Chaired by Jim O'Neill. Report commissioned by the UK Prime Minister.

Shiffman, J. (2014). Knowledge, moral claims and the exercise of power in global health. *International Journal of Health Policy and Management 3*(6), 297–299.

Somekh, B. (2006). *Action research: A methodology for change and development.* Maidenhead: Open University Press.

United Nations. (1987). *Report of the world commission on environment and development: Our common future.*

Appendix: Microbiology Testing at FPRRH

Sample Acquisition

For clinical isolates, the responsibility of diagnosing patients and requesting/taking the samples remained with local healthcare workers. Samples assessed in the hospital laboratory included wound and blood samples. Anonymised isolates were passed on for further investigation as part of this study.

Collection of blood sample: Blood (8–10 ml) was withdrawn from the patient, after the proposed area of skin had been sterilised with ethanol. The blood was contained in a standard BACTEC blood culture bottle, which was transported to the on-site microbiology laboratory where it was cultured in the automated BD BACTEC FX40 machine (37 °C, 5% CO_2) for up to 72 hours to assess any bacterial growth. At the end of the time limit, if no growth was detected, the blood samples were discarded as negative. If growth was present as indicated by the BACTEC FX40 machine, then the bacteria were isolated by streaking onto sheep blood agar plates (BAP) (Sigma-Aldrich) and identified as noted below in isolate confirmation.

Collection of wound samples: A sterile cotton swab was used to gently swab the area of the wound. The swab was transported to the laboratory where it was streaked on to BAPs. The plates were incubated for up to 48 hours and assessed for any growth as noted below in isolate confirmation.

Community isolates were collected from healthcare workers and members of the public through the use of hand swabs.

© The Editor(s) (if applicable) and The Author(s) 2020   179
L. Ackers et al., *Anti-Microbial Resistance in Global Perspective*,
https://doi.org/10.1007/978-3-030-62662-4

Hand swabs: Sterile cotton swabs were first moistened in phosphate buffered saline (PBS) (LabM), then used to swab in between the fingers and across the fingertips of participants. The end of the cotton swabs was broken off and stored in 1ml of PBS for transportation to the laboratory (within 6 hours). All swabs were then streaked on mannitol salt agar (MSA)(LabM) and incubated for up to 48 hours at 37 °C to assess for any growth.

Isolate Confirmation

Identification of Clinical Isolates

Bacterial identification: Positive cultures were identified by picking single colonies from BAPs and performing a series of tests. First, single colonies were sub-cultured onto three different agar plates: sheep blood agar (BAP), chocolate agar (CHOC) (Sigma-Aldrich) and MacConkey agar (MAC). These plates were incubated for 48 hours (35–37 °C, 5% CO_2). Colony colour and morphologies were recorded and assessed by Gram's staining.

Oxidase test: A loop of bacteria was added to the test disk; the colour would change to a dark blue for positive samples and would have no colour change for negative samples.

API strips: Analytical profile index strips contain multiple wells each with different chemicals. A bacterial suspension was added to each well and the results compared to an online database to give a preliminary identification

Catalase test: A small amount of 3% H_2O_2 and fresh bacterial colony were mixed on a glass slide. A positive result was indicated by the production of bubbles of oxygen.

Staphaurex agglutination test: The test reagent was added onto the reaction card followed by the addition of bacterial culture. The mixture was emulsified, and the reaction card gently rotated for up to 20 seconds. Positive samples showed signs of agglutination.

PYR test: Pyrrolidonyl Aminopeptidase test is used for the detection of pyrrolidonyl arylamidase. 5–10 colonies of bacteria were added to the test disk and incubated for 1 minute. 1 drop of N-dimethylaminocinnamaldehyde was added, with positive samples showing the development of a red colour within 1–2 minutes.

Antibiotic Resistance Profiling

The antibiotic sensitivity profile of each bacterial isolate was tested using standard disk diffusion assays following EUCAST guidelines (http:// www.eucast.org/ast_of_bacteria/disk_diffusion_methodology/).   A suspension of each bacterial isolate was prepared in sterile PBS to McFarland Standard 0.5 and then swabbed for semi-confluent growth onto Mueller Hinton agar (LabM). Antibiotic disks (Oxoid) were added to the plate and incubated for 24 hours. Following incubation, the diameter of zones of inhibition was measured and compared to EUCAST breakpoint tables to assess whether the bacteria were resistant. The antibiotics that were assessed included: ampicillin (AMP), ceftriaxone (CEF), chloramphenicol (CMP), ciprofloxacin (CIP), gentamicin (CN), cefoxitin (FOX), cotrimoxazole (COT), erythromycin (E), tetracycline (TE) and ceftriaxone (CRO).

References

Ackers, H. L., Ackers-Johnson, G., Seekles, M., & Opio, S. (2020). Opportunities and challenges for improving anti-microbial stewardship in low- and middle-income countries: Lessons learnt from the maternal sepsis intervention in Western Uganda. *Antibiotics, 9,* 315. https://doi.org/10.3390/antibiotics9060315.

Ackers, H. L., & Ackers-Johnson, J. (2016). *Mobile professional voluntarism and international development: Killing me softly?* Palgrave PIVOT. http://link.springer.com/book/10.1057%2F978-1-137-55833-6.

Ackers, H. L., Ackers-Johnson, J., Tyler, N., & A. Chatwin, J., (2017). *Healthcare, frugal innovation, and professional voluntarism: A cost-benefit analysis.* Cham: Palgrave PIVOT.

Ackers, H. L., Ioannou, E., & Ackers-Johnson, J. (2016). The impact of delays on maternal and neonatal outcomes in Ugandan public health facilities: The role of absenteeism. *Health Policy and Planning, 31,* 1152–1161.

Ackers-Johnson, G. (2020). *Comparing the antimicrobial diversity of Staphylococcus aureus strains isolated from clinical cases of infection and those found as a commensal organism in Fort Portal, Uganda and further investigating the potential mechanisms of resistance present* (PhD research on-going).

Ahmed, A., Ackers-Johnson, J., & Ackers, H. L. (2017). *The ethics of healthcare education placements in low-income countries: First do no harm?* Cham: Palgrave PIVOT.

Allegranzi, B., Nejad, S. B., Combescure, C., Graafmans, W., Donaldson, L., & Pittet, D. (2011). Burden of endemic health-care-associated infection in developing countries: Systematic review and meta-analysis. *Lancet, 377,* 228–241.

© The Editor(s) (if applicable) and The Author(s) 2020
L. Ackers et al., *Anti-Microbial Resistance in Global Perspective,*
https://doi.org/10.1007/978-3-030-62662-4

Allegranzi, B., Aiken, A. M., Zeynep, K. N., Nthumba, P., Barasa, J., Okumu, G., et al. (2018, May). A multimodal infection control and patient safety intervention to reduce surgical site infections in Africa: A multicentre, before-after, cohort study. *The Lancet Infectious Diseases, 18*(5), 507–515. https://doi.org/10.1016/S1473-3099(18)30107-5.

Atwal, A., & Caldwell, K. (2006). Nurse's perceptions of multidisciplinary teamwork in acute health care. *International Journal of Nursing Practice, 12*(6), 359–365.

Baine, S. O., Kasangaki, A., & Baine, E. M. M. (2018). Task shifting in health service delivery from a decision and policy makers' perspective: A case of Uganda. *Human Resources for Health, 16,* 20. https://doi.org/10.1186/s12960-018-0282-z.

Bates, P. (2014). *Context is everything.* Perspectives on Context. London: Health Foundation.

Brink, A. J., van de Bergh, D., Mendelson, M., & Richards, G. A. (2016). Passing the baton to pharmacists and nurses: New models of antibiotic stewardship for South Africa. *South African Medical Journal, 106*(10), 947–948.

Broom, A., Broom, J., & Kirby, E. (2014). Cultures of resistance? A Bourdieusian analysis of doctors' antibiotic prescribing. *Social Science and Medicine, 110,* 81–88.

Bua, J., Mukanga D., Lwanga M., & Nabiwemba, E. (2015). Risk factors and practices contributing to newborn sepsis in a rural district of Eastern Uganda: A cross sectional study. *BMC Res Notes, 8,* 339. Published online 2015, August 9. https://doi.org/10.1186/s13104-015-1308-4.

Cox, J. A., Vlieghe, E., Mendelson, M., Wertheim, H., Ndegwa, L., Villegas, M. V., et al. (2017, November 1). Antibiotic stewardship in low- and middle-income countries: The same but different? *Clinical Microbiology and Infection,* 812–818. https://doi.org/10.1016/j.cmi.2017.07.010.

Cronk, R., & Bartram, J. (2018, April). Environmental conditions in health care facilities in low- and middle-income countries: Coverage and inequalities. *International Journal of Hygiene and Environmental Health, 221*(3), 409–422.

Denyer Willis, L., & Chandler, C. (2018). Anthropology's contribution to AMR control. *Investment and Society,* 104–108. http://resistancecontrol.info/wp-content/uploads/2018/05/104-08-chandler.pdf.

Denyer Willis, L., & Chandler, C. (2019). Quick fix for care, productivity, hygiene and inequality: Reframing the entrenched problem of antibiotic overuse. *BMJ Global Health, 4*(4), e001590. https://doi.org/10.1136/bmjgh-2019-001590.

Dyar, O. J., Huttner, B., Schouten, J., Pulcini, C., & ESGAP (ESCMID Study Group for Antimicrobial stewardshiP). (2017). What is antimicrobial stewardship? *Clinical Microbiology and Infection, 23*(11), 793–798. https://doi.org/10.1016/j.cmi.2017.08.026.

Filippi, V., Ronsmans, C., Gohou, V., Goufodji, S., Lardi, M., Sahel, A., et al. (2005). Maternal wards or emergency obstetric rooms? Incidence of near-miss events in African hospitals. *Acta Obstet Gynecol Scand, 84*, 11–16.

Gardner, A. L., Shunk, R., Dulay, M., Strewler, A., & O'Brien, B. (2018, September). Huddling for high performance teams. *Federal Practitioner*, 16–22. https://www.ncbi.nlm.nih.gov/pmc/articles/PMC6366795/.

Gould, S. J. (1981). *The mismeasure of man*. New York: Norton.

Government of Uganda. (2018). *Antimicrobial Resistance National Action Plan 2018–2023*.

Halbfinger, D. M. (2020, April 21). Hospitals in Israel let relatives say goodbye to loved ones up close. *New York Times*.

Hantrais, L. (2009). *International comparative research: Theory, methods and practice*. London: Palgrave Macmillan.

Harding, S. (1991). *Whose science? Whose knowledge? Thinking from women's lives*. Ithaca, NY: Cornell University.

Harding, S. (2015). *Objectivity and diversity. Another logic of scientific research*. London: University of Chicago Press.

Hoonhout, L., de Bruijne, M. C., Wagner, C., Zegers, M., Waaijman, R., Spreeuwenberg, P., et al. (2009). Direct medical costs of adverse events in Dutch hospitals. *BMC Health Services Research, 9*, 27. https://doi.org/10.1186/1472-6963-9-27.

Hsieh, Y.-H., Lui, J., Tzeng, Y.-H., & Wu, J. (2014). Impact of visitors and hospital staff on nosocomial transmission and spread to community. *Journal of Theoretical Biology, 356*, 20–29.

Hsu, Y.-C., Liu, Y.-A., Lin, M.-H., Lee, H.-W., Chen, T.-J., Chou, L.-F., et al. (2020). Visiting policies of hospice wards during the COVID-19 pandemic: An environmental scan in Taiwan. *International Journal of Environmental Research and Public Health, 17*, 2857.

Kluytmans, J., van Belkum, A., & Verbrugh, H. (1997). Nasal carriage of Staphylococcus aureus: Epidemiology, underlying mechanisms, and associated risks. *Clinical Microbiology Reviews, 10*(3), 505–520.

Kongnyuy, E. J., Mlava, G., & van den Broek, N. (2009). Facility-based maternal death review in three districts in the central region of Malawi: An analysis of causes and characteristics of maternal deaths. *Women's Health Issues, 19*(1), 14–20.

Maina, M., Tosas-Auguet, O., McKnight, J., Zosi, M., Kimemia, G., Mwaniki, P., et al. (2019). Evaluating the foundations that help avert antimicrobial resistance: Performance of essential water sanitation and hygiene functions in hospitals and requirements for action in Kenya. *PLoS ONE, 14*(10), e0222922. https://doi.org/10.1371/journal.pone.0222922.

Mawalla, B., Mshana, S. E., Chalya, P. L., Imirzalioglu, C., & Mahalu, W. (2011). Predictors of surgical site infections among patients undergoing major surgery at Bugando Medical Centre in Northwestern Tanzania. *BMC Surgery, 11*, 21.

Mbabazi, W. (2018). *Assessing compliance with infection prevention and control guidelines among various cadres of healthcare workers at a regional referral hospital in Uganda* (Unpublished MSc thesis).

McCormack, B. (2015). Action research for the implementation of complex interventions. In D. A. Richards & I. R. Hallberg (Eds.), *Complex interventions in health: An overview of research methods* (pp. 300–311). London: Routledge.

Meyer, J. (1993). New paradigm research in practice: The trials and tribulations of action research. *Journal of Advanced Nursing, 18,* 1066–1072.

Meyer, J. (2000). Using qualitative methods in health-related action research. *British Medical Journal, 320,* 178–181.

Michie, S., van Stralen, M. M., & West, R. (2011). The behaviour change wheel: A new method for characterising and designing behaviour change interventions. *Implementation Science, 6,* 42. https://doi.org/10.1186/1748-5908-6-42.

Ministry of Health, Uganda. (2015). *Uganda National Medicines Policy.*

Ministry of Health, Uganda. (2019, September). *The National Annual Maternal and Perinatal Death Surveillance and Response (MPDSR) Report FY 2018/2019.* Ministry of Health Uganda.

Monistrol, O., Calbo, E., Riera, M., Nicolas, C., Font, R., Freixas, N., et al. (2011). Impact of a hand hygiene educational programme on hospital-acquired infections in medical wards. *Clinical Microbiology and Infection, 18,* 1212–1218.

Moore, F. G., Audrey, S., Barker, M., Bond, L., Bonell, C., Hardeman, W., et al. (2015). Process evaluation of complex interventions: Medical Research Council Guidance. *BMJ, 350.* https://doi.org/10.1136/bmj.h1258.

Moyo, D. (2010). *Dead aid: Why aid is not working and how there is another way for Africa.* London: Penguin.

Mutale, W., Balabanova, D., Chintu, N., Mwanamwenge, M. T., & Ayles, H. (2016). Application of system thinking concepts in health system strengthening in low-income settings: A proposed conceptual framework for the evaluation of a complex health system intervention: The case of the BHOMA intervention in Zambia. *Journal of Evaluation in Clinical Practice, 22*(1), 112–121. https://doi.org/10.1111/jep.12160.

Ngonzi, J., Tornes, Y. F., Mukasa, P. K., Salongo, W., Kabakyenga, J., Sezalio, M., et al. (2016). Puerperal Sepsis, the leading cause of maternal deaths at a Tertiary University Teaching Hospital in Uganda. *BMC Pregnancy and Childbirth, 16,* 207.

Oakley, A. (1974). *The sociology of housework.* Oxford: Blackwell.

O'Neill, J. (2016). *The Review on Antimicrobial Resistance. 2016. Tackling drug-resistant infections globally: Final report and recommendations.* The Review on Antimicrobial Resistance, Chaired by Jim O'Neill. Report commissioned by the UK Prime Minister.

Rajkotia, Y. (2018). Beware of the success cartel: A plea for rational progress in global health. *BMJ Global Health, 3*(6), e001197.

Reddy, E. A., Shaw, A. V., & Crump, J. A. (2010). Community-acquired bloodstream infections in Africa: A systematic review and meta-analysis. *Lancet Infectious Diseases, 10*(6), 417–432. https://doi.org/10.1016/S1473-309 9(10)70072-4.

Reinhart, K., Damiles, R., Kisson, N., Machado, F. R., Schachter, R. D., & Finfer, S. (2017). Recognizing sepsis as a global health priority—A WHO resolution. *New England Journal of Medicine, 377,* 414–417.

Richards, D. A., & Hallberg, I. R. (2015). *Complex interventions in health.* London: Routledge.

Rochford, C., Sridhar, D., Woods, N., Saleh, Z., Hartenstein, L., Ahlawat, H., et al. (2018). Global governance of antimicrobial resistance. *Lancet, 391*(10134), 1976–1978.

Rowe, S. Y., Peters, D. H., Holloway, K. A., Chalker, J., Ross-Degnan, D., & Rowe, A. K. (2019). A systematic review of the effectiveness of strategies to improve health care provider performance in low- and middle-income countries: Methods and descriptive results. *PLoS ONE, 14*(5), e0217617.

Saha, A., & Alleyne, G. (2018). Recognising noncommunicable disease as a global health security threat. *Bulletin of the World Health Organisation, 96,* 792–793.

Seni, J., Najjuka, C. F., Kateete, D. P., Makobore, P., Joloba, M. L., Kajumbula, H., et al. (2013). Antimicrobial resistance in hospitalised surgical patients: A silently emerging public health concern in Uganda. *BMC Research Notes, 6,* 298.

Shafquat, Y., Jabeen, K., Farooqi, J., Mehmood, K., Irfan, S., Hasan, R., et al. (2019). Antimicrobial susceptibility against metronidazole and carbapenem in clinical anaerobic isolates from Pakistan. *Antimicrobial Resistance and Infection Control, 8,* 99. Published 2019 June 14. https://doi.org/10.1186/s13 756-019-0549-8.

Shiffman, J. (2014). Knowledge, moral claims and the exercise of power in global health. *International Journal of Health Policy and Management, 3*(6), 297–299.

Slawomirski, L., Auraaen, A., & Klazinga, N. (2017). *The economics of patient safety.* OECD.

Smith, A. (2018). Metronidazole resistance: A hidden epidemic? *British Dental Journal, 224*(6), 403–404. https://doi.org/10.1038/sj.bdj.2018.221.

Somekh, B. (2006). *Action research: A methodology for change and development.* Maidenhead: Open University Press.

Storeng, K. T., & Palmer, J. (2019). When ethics and politics collide in donor-funded global health research. *Lancet, 394,* 184–186.

Tong, S. Y., Davis, J. S., Eichenberger, E., Holland, T. L., & Fowler, V. G. (2015). Staphylococcus aureus infections: Epidemiology, pathophysiology, clinical manifestations, and management. *Clinical Microbiology Reviews, 28*(3), 603–661.

Tweheyo, R., Reed, C., Campbell, S., Davies, L., & Daker-White, G. (2019). 'I have no love for such people, because they leave us to suffer': A qualitative study of health workers' responses and institutional adaptations to absenteeism in rural Uganda. *BMJ Global Health, 4,* e001376. https://doi.org/10.1136/bmjgh-2018-001376.

United Nations. (1987). *Report of the World Commission on Environment and Development: Our common future.*

Uganda National Academy of Sciences. (2015). *Antimicrobial resistance in Uganda: Situation analysis and recommendations.* Centre for Disease Dynamics, Economics & Policy. ISBN: 978-9970-424-10-8.

Vogel, I. (2012). *Review of the use of 'Theory of Change' in international development.* Review Report Department for International Development https://assets.publishing.service.gov.uk/media/57a08a5ded91 5d3cfd00071a/DFID_ToC_Review_VogelV7.pdf.

Weber, N., Patrick, M., Hayter, A., Martinson, A. L., & Gelting, R. (2018). A conceptual evaluation framework for the water and sanitation for health facility improvement tool (WASH FIT). *Journal of Water, Sanitation and Hygiene for Development, 9*(2). https://doi.org/10.2166/washdev.2019.090.

Wee, L. E., Conceicao, E. P., Sim, X., Y. J., Aung M. K., Tan, K. Y., Wong, H. M., et al. (2020). Minimizing intra-hospital transmission of COVID-19: The role of social distancing, *Journal of Hospital Infection, 105,* 113–115.

Welsh, J. (2019). *Providing an evidence base for antibiotic stewardship for midwives in the Kabarole District of Uganda: A modified action research study* (Unpublished PhD thesis).

World Bank. (2017). *Drug-resistant infections: A threat to our economic future.* Washington, DC: World Bank.

World Health Organisation. (2007). *Task shifting to tackle health worker shortages.* Retrieved from: http://www.who.int/healthsystems/task_shifting_book let.pdf. Accessed 4 March 2017.

World Health Organisation. (2008). *Task shifting. Global recommendations and guidelines.* Geneva: World Health Organisation.

World Health Organisation. (2012). *Optimizing health worker roles to improve access to key maternal and new-born health interventions through task shifting.* Retrieved from: http://apps.who.int/iris/bitstream/handle/10665/77764/ 9789241504843_eng.pdf;jsessionid=876E08C843C919EA6CE890EB370 BEAB8?sequence=1. Accessed 1 February 2018.

World Health Organisation. (2019). *Patient safety fact file.*

Wright-Mills, C. (1959). *The sociological imagination.* New York: Oxford University Press.

Zulfiuqar, B., Salam, A., Firoz, M., Fatima, H., & Aziz, S. (2013). Effects of inflow of inpatients attendants at a tertiary care hospital—A study at civil hospital Karachi. *Journal of Pakistan Medical Association, 63*(1), 143–147.

INDEX